Junbai Li, Qiang He, and Xuehai Yan

Molecular Assembly of Biomimetic Systems

Related Titles

Öchsner, A., Ahmed, W. (eds.)

Biomechanics of Hard Tissues

Modeling, Testing, and Materials

2010
ISBN: 978-3-527-32431-6

Kumar, C. S. S. R. (ed.)

Biomimetic and Bioinspired Nanomaterials

Series: Nanomaterials for the Life Sciences (Volume 7)

2010
ISBN: 978-3-527-32167-4

Junbai Li, Qiang He, and Xuehai Yan

Molecular Assembly of Biomimetic Systems

WILEY-VCH Verlag GmbH & Co. KGaA

The Authors

Prof. Dr. Junbai Li
Chinese Academy of Sciences
Institute of Chemistry
Zhongguancun North 1st Str. 2
Beijing 100080
People's Republic of China

Dr. Qiang He
Chinese Academy of Sciences
Institute of Chemistry
Zhongguancun North 1st Str. 2
Beijing 100080
People's Republic of China

Dr. Xuehai Yan
Chinese Academy of Sciences
Institute of Chemistry
Zhongguancun North 1st Str. 2
Beijing 100080
People's Republic of China

All books published by **Wiley-VCH** are carefully produced. Nevertheless, authors, editors, and publisher do not warrant the information contained in these books, including this book, to be free of errors. Readers are advised to keep in mind that statements, data, illustrations, procedural details or other items may inadvertently be inaccurate.

Library of Congress Card No.: applied for

British Library Cataloguing-in-Publication Data
A catalogue record for this book is available from the British Library.

Bibliographic information published by the Deutsche Nationalbibliothek
The Deutsche Nationalbibliothek lists this publication in the Deutsche Nationalbibliografie; detailed bibliographic data are available on the Internet at <http://dnb.d-nb.de>.

© 2011 Wiley-VCH Verlag & Co. KGaA, Boschstr. 12, 69469 Weinheim, Germany

All rights reserved (including those of translation into other languages). No part of this book may be reproduced in any form – by photoprinting, microfilm, or any other means – nor transmitted or translated into a machine language without written permission from the publishers. Registered names, trademarks, etc. used in this book, even when not specifically marked as such, are not to be considered unprotected by law.

Typesetting   Toppan Best-set Premedia Ltd., Hong Kong
Printing and Binding   betz-druck GmbH, Darmstadt
Cover Design   Formgeber, Eppelheim

Printed in the Federal Republic of Germany
Printed on acid-free paper

ISBN: 978-3-527-32542-9

Contents

Preface *IX*

Introduction *1*
Biomimetic Membranes *2*
Layer-by-Layer Assembly of Biomimetic Microcapsules *2*
F_oF_1-ATP Synthase-Based Active Biomimetic Systems *3*
Kinesin–Microtubule-Driven Active Biomimetic Systems *3*
Biomimetic Interface *4*
Peptide-Based Biomimetic Materials *4*

1 Biomimetic Membranes *7*
1.1 Introduction *7*
1.2 Lipid Monolayers *8*
1.2.1 Phospholipid Monolayers at the Air/Water Interface *8*
1.2.2 Phospholipid Monolayers at the Oil/Water Interface *10*
1.2.3 Interfacial Behavior of Phospholipid Monolayers *10*
1.2.4 Protein Layers at the Oil/Water Interface *15*
1.2.4.1 Kinetics of Protein Adsorption *15*
1.2.4.2 Formation of "Skin-Like" Protein Films on a Curved Interface *16*
1.2.5 Interfacial Behavior of Phospholipid/Protein Composite Layers *17*
1.2.5.1 Dynamic Adsorption and Mechanism *18*
1.2.5.2 Assembly of "Skin-Like" Complex Films on a Curved Interface *19*
1.3 Modeling Membrane Hydrolysis *In Vitro* *20*
1.3.1 PLA2 *21*
1.3.2 PLC *23*
1.3.3 PLD *24*
1.4 Polyelectrolyte-Supported Lipid Bilayers *25*
1.4.1 Polyelectrolyte Multilayers on Planar Surfaces *27*
1.4.2 Polyelectrolyte Multilayers on Curved Surfaces *28*
1.5 Conclusions and Perspectives *30*
References *31*

2	**Layer-by-Layer Assembly of Biomimetic Microcapsules** 41
2.1	Introduction 41
2.2	Layer-by-layer Assembly of Polyelectrolyte Multilayer Microcapsules 42
2.2.1	General Aspects 42
2.2.2	Permeation and Mechanical Properties of LbL Microcapsules 45
2.3	Biointerfacing Polyelectrolyte Microcapsules – A Multifunctional Cargo System 47
2.3.1	Lipid Bilayer-Modified Polyelectrolyte Microcapsules 48
2.3.2	Formation of Asymmetric Lipid Bilayers on the Surface of LbL-Assembled Capsules 51
2.3.3	Assembly of Lipid Bilayers on Covalently LbL-Assembled Protein Capsules 53
2.4	Application of Biomimetic Microcapsules 55
2.4.1	Integrating Specific Biofunctionality for Targeting 55
2.4.2	Adsorption of Antibodies on the Surface of Biomimetic Microcapsules 57
2.5	Conclusions and Perspectives 58
	References 58

3	**F_oF_1-ATP Synthase-Based Active Biomimetic Systems** 63
3.1	Introduction 63
3.2	F_oF_1-ATPase – A Rotary Molecular Motor 63
3.2.1	Structure of $H^+F_oF_1$-ATPase 64
3.2.2	Direct Observation of the Rotation of Single ATPase Molecules 67
3.3	Reconstitution of F_oF_1-ATPase in Cellular Mimic Structures 70
3.3.1	F_oF_1-ATPase-incorporated Liposome – A Classical Biomembrane Mimic 71
3.3.1.1	Bacteriorhodopsin uses Light to Pump Protons 71
3.3.1.2	Proton Gradients Produced by Artificial Photosynthetic Reactions 74
3.3.2	ATP Biosynthesis from Biomimetic Microcapsules 76
3.3.2.1	Generation of Proton Gradients in Polymer Capsules by the Change of pH Values 76
3.3.2.2	Proton Gradients in Protein Capsules Supplied by the Oxidative Hydrolysis of Glucoses 78
3.3.2.3	Proton Gradients Generated by GOD Capsules 80
3.3.3	Reassembly of F_oF_1-ATPase in Polymersomes 82
3.4	Conclusions and Perspectives 85
	References 85

4	**Kinesin–Microtubule-Driven Active Biomimetic Systems** 91
4.1	Introduction 91
4.2	Kinesin–Microtubule Active Transport Systems 92
4.3	Active Biomimetic Systems Based on the Kinesin–Microtubule Complex 93

4.3.1	Bead Geometry	93
4.3.2	Gliding Geometry	94
4.3.3	Transport Direction and Distance of Assembled Systems	95
4.4	Layer-by-Layer Assembled Capsules as Cargo – A Promising Biomimetic System	96
4.4.1	Layer-by-Layer Assembled Hollow Microcapsules	96
4.4.2	Kinesin–Microtubule-Driven Microcapsule Systems	97
4.4.2.1	Fabrication in a Beaded Geometry	97
4.4.2.2	Fabrication in a Gliding Geometry	98
4.5	Conclusions and Perspectives	100
	References	101
5	**Biomimetic Interface**	**103**
5.1	Introduction	103
5.2	Preparation and Characterization of Biomolecule Patterning	104
5.2.1	Electrostatic Immobilization of Proteins for Surface Assays	105
5.2.1.1	Lipid-Modified HSA Patterns for the Targeted Recognition	105
5.2.1.2	Lipid-Modified HSA Patterns for *Escherichia coli* Recognition	107
5.2.2	Covalent Immobilization of Proteins	108
5.2.3	Covalent Immobilization of Lipid Monolayers	110
5.3	Polymer Brush Patterns for Biomedical Application	114
5.3.1	Thermosensitive Polymer Patterns for Cell Adhesion	114
5.3.2	Fabrication of Complex Polymer Brush Gradients	118
5.4	Conclusions and Perspectives	123
	References	124
6	**Peptide-Based Biomimetic Materials**	**129**
6.1	Introduction	129
6.2	Peptides as Building Blocks for the Bottom-up Fabrication of Various Nanostructures	130
6.2.1	Aromatic Dipeptides	130
6.2.1.1	Nanotubes, Nanotube Arrays, and Vesicles	131
6.2.1.2	Nanofibrils and Ribbons	136
6.2.1.3	Nanowires	139
6.2.1.4	Ordered Molecular Chains on Solid Surfaces	140
6.2.2	Lipopeptides	140
6.2.3	Polypeptides	146
6.2.4	Amphiphilic Peptides	150
6.3	Peptide–Inorganic Hybrids	156
6.3.1	Nonspecific Attachment of Inorganic Nanoparticles on Peptide-Based Scaffolds	156
6.3.2	Peptide-Based Biomineralization	160
6.3.3	Adaptive Hybrid Supramolecular Networks	164
6.4	Applications of Peptide Biomimetic Nanomaterials	165
6.4.1	Biological Applications	165

6.4.1.1 Three-Dimensional Cell Culture Scaffolds for Tissue Engineering *165*
6.4.1.2 Delivery of Drugs or Genes *167*
6.4.1.3 Bioimaging *169*
6.4.1.4 Biosensors *170*
6.4.2 Nonbiological Applications *170*
6.5 Conclusions and Perspectives *171*
References *172*

Glossary *183*
Index *187*

Preface

We are frequently asked how much we can learn from nature. In most case, we can get the answers by biomimetics. With the development of nanosciences, biomimetics is staring at the molecular level. This is based on the fact that many bioactive molecules like DNA, lipid, peptide, proteins can self-assemble into well-defined structures and further to a supramolecular architecture while combining with other organic, inorganic or metal oxide compounds. It is therefore considered the promising method to fabricate novel materials. The obvious feature of such biomimetic systems are their artificial structures which can be inspired by biology. A major advantage of these assembled systems is that they keep their biochemical and physical parameters and properties in a controlled manner. Thus the intense interest in this field is clearly evident.

The present book attempts to introduce the aspects and practical techniques of molecular assembly of biomimetic systems, especially, the layer-by-layer assembly, self-assembly, microcontact printing, electron beam lithography and chemical lithography.

We have benefited from many efforts of our co-workers in making this book reality. We sincerely acknowledged them, notably Weixing Song, Zhihua An, Liqin Ge, Cheng Tao. We have to say that we have learned a lots about molecular assembly from Profs. H. Möhwald, H. Rinsdorf, T. Kunitake and Jiacong Shen who have done much of the pioneering work and we are grateful to all of them for their motivated and skillful helps and contributions.

Beijing, November 2010

Junbai Li
Qiang He
Xuehai Yan

Introduction

In nature, biological systems and physiological processes have evolved over millions of years to improve their properties and functions. Biomimetics, simply, is the attempt to mimic these properties and functions. It involves studying structures and mechanisms of tissue formation in the organisms. Using biology as a guide, we can now understand, engineer, and control bioactive molecular interactions, and assemble them into novel systems or materials. The molecular biomimetic approach opens up new avenues for the design and utilization of multifunctional molecular systems with a wide range of applications in nanotechnology and biotechnology. Molecular assembly of biomimetic systems is now regarded as one of the promising methods to fabricate well-defined nanostructures and materials, and its importance is now generally recognized.

Biomimetic systems are artificial structures that are inspired by biology. A major advantage of these systems is that both biochemical and physical parameters can be controlled precisely. Therefore, it is feasible to utilize biomimetic systems as experimental models for guiding research on biological mutation and evolution in organisms. Some bioactive molecules such as peptides, proteins, nucleic acids, and lipids can undergo self-assembly into well-defined structures similar to the assembly in living organs. Biomimetics is not limited to just copying nature because, with the development of modern biology, scientists can directly utilize biological units themselves to construct new types of systems sometimes as hybrid nanostructured materials. In this way, some of the manufacturing difficulties of biomimetics can be avoided. As will be illustrated in this book, natural molecular machines such as motor proteins are integrated into the engineering of active biomimetic systems so that new functionalized systems can be constructed.

This book covers fundamental aspects and practical techniques of the molecular assembly of biomimetic systems; in particular, layer-by-layer (LbL) assembly, self-assembly, microcontact printing, electron beam lithography (EBL), and chemical nanolithography. It also presents an overview of the molecular assembly of biomimetic systems that consists of the following six topics covered in individual chapters.

Molecular Assembly of Biomimetic Systems. Junbai Li, Qiang He, and Xuehai Yan
© 2011 WILEY-VCH Verlag GmbH & Co. KGaA, Weinheim
ISBN: 978-3-527-32542-9

Biomimetic Membranes

Biological membranes are key components in biological systems, forming the natural boundary of cells to separate inner components from outer environment. A number of cell actions and functions relevant to the environment are fulfilled via membrane processes, in most cases depending on the interactions of membrane proteins and carbohydrates. Owing to the complexity of biological membranes, it is very important to design and engineer artificial model membranes to overcome the difficulty of investigating membrane function directly in living cells. In this regard, biomimetic membranes as either a lipid monolayer or lipid bilayers are created by simple artificial strategies, such as spread at air/water or oil/water interfaces, or vesicle fusion. A lipid monolayer fixed at the air/water interface is a popular model membrane for investigation of the hydrolysis process catalyzed by enzymes at the interface. More attention is being paid to lipid bilayers supported on a planar surface or a curved surface because supported membranes, especially cushioned by polymers, are an ideal model to unravel the physical and chemical properties of biomembranes and their contribution to membrane functions.

In Chapter 1, we primarily focus on the fabrication of a lipid monolayer as a simplified model for studying the dynamic adsorption and interfacial behavior as well as membrane hydrolysis process catalyzed by enzymes at the interface. In addition, we briefly summarize the most recent developments and applications on supported lipid bilayers at polyelectrolyte multilayers, including at planar and curved surfaces. The introduction of such a biomimetic membrane will enhance greatly our understanding of the function and property of biological membranes, and will also significantly help to develop advanced characterization tools or techniques for a better understanding of the biological membrane system.

Layer-by-Layer Assembly of Biomimetic Microcapsules

Biomimetic microcapsules are a class of artificial hollow sacs with controllable size and versatile function such as tunable physicochemical properties and permeability. They can be regarded as a promising cell mimic to simulate some functions of cell membranes. Quite different from conventional liposomes, such a biomimetic hollow sac contains a large enough compartment so that the natural environment of membrane-bound proteins can be recreated. The materials making up microcapsules are a variety of polymers that are beneficial for the affinity and embedment of membrane proteins on supported biological membranes. The LbL assembly technique, which was first developed by Decher for the fabrication of ultrathin multilayers, is effective in the preparation of a hollow shell upon colloidal templates. Biogenic microcapsules prepared by the LbL technique are of great interest due to their potential application in medicine, catalysis, cosmetics, and biotechnology. By the conversion of liposomes into lipid bilayers, the coating of active lipid bilayers on polymer microcapsules can readily be achieved. Such lipid-decorated microcapsules can serve as an ideal supported biomimetic membrane system to mimic functions of the cell membrane. This new hybrid system also enables the design and application of new biomimetic structured materials.

In Chapter 2, we describe how LbL-assembled polyelectrolyte microcapsules can be interfaced with biological components such as phospholipid membranes and proteins. LbL assembly has attracted extensive attention for the fabrication of biomimetic microcapsules because it provides engineered features including size, shape, thickness, composition and permeation, and the capability of incorporating different types of biomolecules. The applications of these biomimetic microcapsules in drug delivery, biosensors, and hybrid nanodevices are also addressed.

F_oF_1-ATP Synthase-Based Active Biomimetic Systems

ATP synthase (ATPase) is one of the most popular molecular machines and has been extensively studied. It can act as a rotary motor in the design of novel nanodevices, continuously synthesizing ATP in the artificial environment. Production of ATP is one of the most important chemical reactions in living biology. With regard to the production of ATP, ATPase is the primary enzyme to catalyze the reaction where the generation of ATP from ADP and inorganic phosphate is performed upon the induction of proton gradients. The functionality of ATPase has attracted great interest over the last decade. Many potential applications have been suggested, from the generation of bioenergy to the fabrication of nanodevices. Lipid membranes have been widely used as models for biological membranes and ATPase is particularly selected as a model membrane protein, since it is a major ATP supplier in the cell. As a membrane-bound protein, ATPase can be reconstituted *in vitro* into liposomes via detergent mediation. Nevertheless, the limitations of the size and instability of the assembled liposome complexes result in difficulties in understanding and analyzing the system. Instead, lipid-coated polymer microcapsules exhibit extensive advantages as a biomimetic vehicle having a similar function to liposomes, but controllable in size and robust in structure.

In Chapter 3, we explore how biomimetics can be applied to engineering functional nanomaterials, particularly to assembling ATPase in artificial containers and mimetic cellular systems with cellular processes. Much effort has been focused on assembling ATPase in biomimetic systems so that a complex cellular process can be constructed in a controllable manner. Recently, LbL-assembled microcapsules have proven to be a suitable cellular mimetic structure and have been applied to engineering active biomimetic systems with cellular processes. An added benefit is that these assembled microcapsules can be used as bioenergy containers and thus supply ATP on demand.

Kinesin–Microtubule-Driven Active Biomimetic Systems

Linear motor proteins such as kinesin and myosin can transport cargoes inside cells with both spatial and temporal precision. These linear motor proteins provide the inspiration of the design and build-up of novel biomimetic functional nanomaterials. Kinesins are a family of proteins that can be divided into 14 classes based on sequence similarity and functional properties. Over the past decades, efforts to use linear motor proteins as nanoactuators have made rapid progress. In general, these motor proteins consume chemical energy to power the movement of targeted components into devices engineered at the micro- and nanoscale. The design of such hybrid nanodevices requires suitable synthetic environments and the identification of unique applications. Linear cytoskeletal kinesin motors have dominated the emerging field of protein-powered devices because they are relatively robust and readily available. Tubulin can be commercially purchased, while the motor proteins can be purified from cells or expressed in recombinant bacterial systems and harvested in large quantities.

In Chapter 4, the recent progress of assembling kinesin–microtubule–cargo systems in a synthetic environment is presented. In particular, we discuss the selection, loading, and unloading of cargoes, and also highlight our ongoing work – LbL-assembled microcapsules serving as cargoes driven by kinesin motors.

Biomimetic Interface

Biomimetic interface engineering modifies the interfaces between biological and nonbiological systems to gain valuable insight into the biological interactions at these interfaces. The main advantage of biomimetic interface strategies is the ability to influence biological interactions by modifying the interfaces, while still retaining the vital physical properties and to some extent improving the biocompatibility of the materials. A number of methods or techniques, including optical lithography, nanoimprint lithography, dip-pen nanolithography, and microcontact printing, are available for the engineering or patterning of interfaces. These biologically functionalized interfaces, generally as biomimetic interfaces, have a wide range of applications in biology and nanotechnology (e.g., for drug delivery, biosensors, biochips and medical implants, etc.).

In Chapter 5, we provide a brief overview of the advances in the application of microcontact printing for lipid micropatterning, and EBL for lipid nanopatterning and polymer gradient structures. In particular, a relatively new technique, chemical nanolithography, which is based on radiation-induced changes in organic self-assembled monolayers, is addressed.

Peptide-Based Biomimetic Materials

Self-assembly of biological building blocks has attracted increasing attention due to their versatility for bottom-up fabrication, biocompatibility, and biodegradability, with a wide range of application in biology and nanotechnology. Many biomolecules including peptides and proteins can interact and self-assemble into highly ordered supramolecular architectures with functionality. In the self-assembly process the precise control of supramolecular architectures is achieved through synergistic effects of some weak noncovalent interactions such as hydrogen bonds, electrostatic interactions, π–π stacking, hydrophobic forces, nonspecific Van der Waals forces, chiral dipole–dipole interactions, and so on. Although these forces are individually weak, when combined as a whole, they govern self-assembly of molecular building blocks into superior and ordered superstructures. Self-assembly is ubiquitous in nature. By learning from nature one can purposefully devise and synthesize artificial building blocks amenable to self-assembly into superstructures by cooperative interactions of weak noncovalent interactions. Notably, peptides composed of several to dozens of amino acids have been of great interest in the creation of biomimetic or bioinspired nanostructured materials owing to their structural simplicity and tunability, functional versatility, cost-effectiveness, and widespread applications.

In Chapter 6, we first focus on the fabrication of peptide-based nanostructural materials from synthetic building blocks such as lipopeptides, polypeptides, amphiphilic peptides and, particularly, diphenylalanine-based peptides derived from Alzheimer's β-amyloid polypeptide. In addition, we present the experimental results and progress in the integration of peptide biomaterials with functional inorganic components for creating multifunctional materials. We then discuss the potential applications of such assembled peptide-based materials in biological and nonbiological areas, including tissue engineering, gene or drug delivery, bioimaging and biosensors, as well as functional templates for nanofabrication.

1
Biomimetic Membranes

1.1
Introduction

The cell membrane is a selectively permeable lipid bilayer that is a basic structural unit in all cells. It is composed of a variety of biological molecules such as lipids, and proteins and lipopolysaccharides are attached to the membrane surface. A vast array of cellular processes, including cell adhesion, ion channel conductance, and cell signaling, are performed at such biological interfaces [1]. The cell membrane is a natural barrier on a cell that separates the intracellular components from the extracellular environment. The biological functions associated with membranes involve a number of different molecular species. Both the lipid and protein compositions of membranes are primarily responsible for membrane function as well as structure. Membrane proteins embedded in the cell membrane are of particular importance in adjusting and controlling the delivery of substances across the membrane and acting as molecular signals that allow cells to communicate with each other. Additionally, protein function can be influenced by the lipid matrix that surrounds it. Understanding the function and structure of cell membranes remains a critical challenge. Altogether, the biological membrane has proved to be crucial for cell survival. The phospholipid bilayer functions in compartmentalization, protection, and osmoregulation, and the proteins have a wide range of functions including molecular recognition, transport of substances, and metabolic processes [2]. Thus, model membranes have historically been indispensable for the development of our understanding of biological membranes. Such biomimetic systems allow us to study the individual features of these highly complex structures.

Lipid membranes are one of most important self-assembled structures in nature. Lipid monolayers or bilayers prepared artificially provide excellent model systems for studying the surface chemistry of biological membranes. To some extent, such a type of membrane is a biomimetic structure. Their use can help us to understand the structure and function of membranes, and the relationship between them through simplifying experiment processes, reducing complexity in

Molecular Assembly of Biomimetic Systems. Junbai Li, Qiang He, and Xuehai Yan
© 2011 WILEY-VCH Verlag GmbH & Co. KGaA, Weinheim
ISBN: 978-3-527-32542-9

data interpretation, and improving greater experimental control [3]. It is well known that biological membranes consist of lipid bilayers where other components such proteins and enzymes are able to penetrate. To understand the biological processes of membranes at a molecular level, a lipid monolayer is often considered as a simplified model of biomembranes (e.g., the self-assembled lipid monolayer at the air/water interface has been used as a model membrane for investigating the hydrolysis process catalyzed by enzymes at the interface) [4, 5]. Supported lipid bilayers [6, 7] have become more and more important biomimetic materials, and are popular models of cell membranes for potential application in biology and nanotechnology. The construction of the model membrane is significantly advantageous to understand the function of biological membranes *in vitro* (e.g., investigating protein ligand–receptor interactions [8–11], cellular signaling events [12–14], biological sensing, and transport roles of biological membranes [15–17]). During the past decade a large number of solid-supported lipid membranes have been developed, including inorganic surfaces, functionalized monolayers, polymers, and others. With the increasing demand for supported membranes, polyelectrolytes are emerging as popular choices for cushioning materials. Usually, polyelectrolytes are self-assembled to a variety of substrates by means of the layer-by-layer (LbL) deposition method, providing potential control over the resulting cushion film thickness, porosity, polarity gradient, and so on. Thus, it is possible to provide a new hydrated environment for the lipid membrane that serves as a versatile biomimetic membrane in which function and property can be varied on demand.

In this chapter, we focus primarily on the fabrication of lipid monolayers as a simplified model for studying their dynamic adsorption and interfacial behavior, as well as membrane hydrolysis processes catalyzed by enzymes at the interface. Additionally, we briefly summarize the most important new developments and applications on the supported lipid bilayers at the polyelectrolyte multilayers, including at planar and curved surfaces. The introduction of such a biomimetic membrane will greatly enhance our understanding to the function and property of biological membranes, and will also help to develop advanced characterization tools or techniques for better investigation of biological membrane systems.

1.2
Lipid Monolayers

1.2.1
Phospholipid Monolayers at the Air/Water Interface

With increasing interest in biological membranes, phospholipid monolayers have been fabricated as a simple model membrane for studying the corresponding interfacial behavior. Lipid monolayers at the air/water interface can be easily formed by spreading an organic solution of phospholipid on the surface of water [18, 19]. The hydrophilic headgroups of the phospholipid face towards the water while the hydrophobic alkyl chains are exposed to the air. The self-assembly nature of the phospholipid molecule leads to the formation of a two-dimensional lipid phase.

The phase state of this monolayer is pertinent to the phospholipid concentration at interfaces (molecules/area). We can obtain a two-dimensional phase diagram when compressing the phospholipid monolayer at a constant temperature (i.e., decrease the area available to the phospholipid). Surface pressure (π) as a function of the molecular area (A) (i.e., π–A isotherm) may give integral information on the lipid-phase transition [20–22]. For instance, the π–A isotherm of the L-α-dipalmitoylphosphatidylcholine (L-DPPC) monolayer undergoes almost all of the possible phase states of the insoluble monolayer. As shown in Figure 1.1, with the increase of surface pressure, different molecule packing patterns of L-DPPC in a monolayer will appear. At the beginning, the monolayer is in the gas-phase state and the molecular arrangement is completely disordered. When the monolayer is compressed by a lateral surface force, the molecules become closer and subsequently go through a phase change from gas to the liquid-expanded phase, followed by the coexistence region of the liquid-expanded and the liquid-condensed phase. If the given pressure is large enough, the stacking phase state of the monolayer is also likely as the solid state, in which the molecules are closely packed and orderly oriented along a certain direction. However, when the given pressure is over a

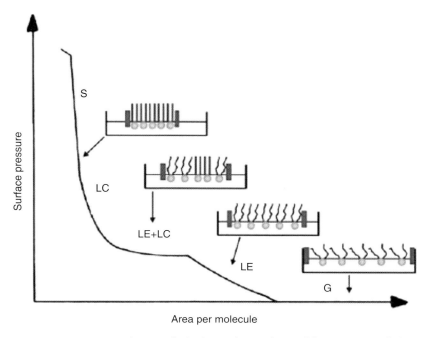

Figure 1.1 Pressure/area diagram of a lipid monolayer: scheme of the compression behavior and packing in the gas (G), liquid-expanded (LE), liquid-compressed (LC), and solid (S) phases. (Reprinted with permission from [5]. © 2007, Elsevier.)

certain limited value, the monolayer will collapse. It should be pointed out that the shape of the isotherms depends greatly on the structure of the phospholipid itself. Therefore, the π–A isotherm of lipid molecules is an important parameter that allows us to acquire information on their orientation, mobility, and interactions with compounds dissolved in the subphase.

1.2.2
Phospholipid Monolayers at the Oil/Water Interface

Phospholipids are also effective natural surface-active substances just like a surfactant and act as good stabilizers of emulsion [23–26]. A study of the adsorption behavior and interfacial rheology of such systems is important to obtain knowledge and understanding of interfacial activity and system stability [27–34]. Unlike lipid monolayers at the air/water interface, whose surface tension and dilational parameters can be measured by the Langmuir method, it is experimentally difficulty to modify this method to study the liquid/liquid interface [35]. A powerful method to obtain information on the thermodynamic and dynamic properties of lipid monolayers at the liquid/liquid interface is the drop profile analysis tensiometry (PAT) technique, which was introduced by Neumann *et al.* in the early 1990s [36]. In this case, the volume of a pendent drop is increased or reduced, thus resulting in expanding or compressing the surface of the drop. Using the PAT technique, the surface tension becomes accessible as a function of the drop surface area. When the changes of the surface tension follow a harmonic function it is easiest to extract the surface dilation parameters. At present, a large number of commercial instruments are available that use this quite efficient methodology for rheological measurements at gas/liquid and liquid/liquid interfaces.

1.2.3
Interfacial Behavior of Phospholipid Monolayers

Phospholipid monolayers self-assembled at the air/water interface have been widely studied, aiding as precursors of well-structured organic films and models of biological membranes. Actually, biomimetic monolayers fabricated at the oil/water interface also have vital implications for gaining insights into the interfacial behavior of amphiphilic phospholipid molecules. PAT is an effective technique to determine the interfacial tension, contact angle, drop surface area, and drop volume based on the drop profile in a variety of situations [36–39]. It can also be used to study these parameters depending on time, temperature, and pressure [40, 41]. In addition, PAT is also applicable for investigating the relaxation of surfactant adsorption layers [42]. This relaxation is performed by sudden changes of the drop volume and hence the surface area of a pendent drop while images of the drop profiles are taken [43]. Therefore, PAT is a very versatile methodology in investigating the interfacial behavior of amphiphiles. The basic experimental procedure can be found in Figure 1.2. Next, DPPC and dimyristoylphosphatidylethanolamine (DMPE) are selected as model phospholipid molecules to elu-

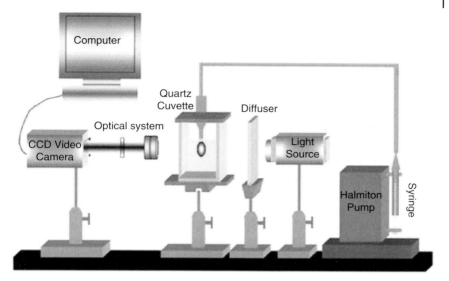

Figure 1.2 Experimental setup for drop profile analysis. The pure chloroform or lipid/chloroform solution was pumped into a pipeline by a Hamilton pump to the end of a capillary for producing a pendent drop. The drop was immersed into the protein buffer solution. The drop size can be controlled precisely by a Hamilton pump via a computer. A few drops were eliminated and left to the bottom of the cuvette in order to create a saturated environment and form a fresh drop. The capture and analysis of the pendent drop was based on the ADSA (axisymmetric drop shape analysis) commercial software of. The forming and detaching times of the drop were accurately recorded by a video recorder.

cidate their dynamic adsorption behavior at the oil/water interface by the PAT technique.

PAT can function as a film balance to measure the π–A isotherm at the liquid/liquid interface [44]. Compared with the conventional Langmuir trough method, surface tension and area are direct output data so that continuous measurement of surface tension as a function of time is possible by changing the drop volume. In addition, when PAT is applied as a film balance at the liquid/liquid interface, a lower quantity of spreading materials is required. Furthermore, PAT supplies uniform concentration, temperature, and pressure of the system. Figure 1.3 shows an example of a π–A isotherm measured for a DPPC film at 20 °C by using PAT and Langmuir–Blodgett (LB) balance at the chloroform/water interface [41]. Obviously, the isotherms measured by the two methods show good agreement.

There are a variety of factors able to influence the results of dynamic interfacial tension, including different molecular structures, pH value, ionic strength, temperature, concentration, and so on. For the phospholipids, their dynamic interfacial tension is subjected to the concentration and different headgroups and hydrophobic chains of lipids when studied at the pure water/chloroform interface [45, 46]. Concentrations below 0.001 mM for DPPC and 0.002 mM for DMPE

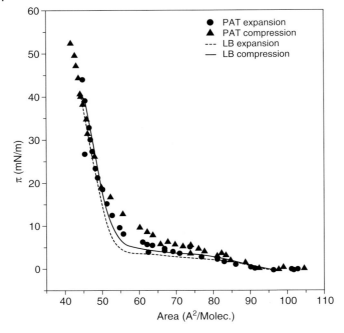

Figure 1.3 Surface pressure–area isotherm of a DPPC film measured at 20 °C by the PAT method (PAT is used as a Langmuir balance). (Reprinted with permission from [41]. © 1996, Elsevier.)

exhibit no measurable change in the interfacial tension; at concentrations above 0.002 (DPPC) and 0.003 mM (DMPE), the interfacial tension changes continuously toward the equilibrium state. For higher concentrations (i.e., 0.02–0.05 mM for DPPC and 0.015–0.1 mM for DMPE), the interfacial tensions reach equilibrium values within 60 s. Johnson and Saunders [47] studied the interfacial tension of a series of phosphatidylcholines, $(C_{12})_2PC$, $(C_{14})_2PC$, and $(C_{16})_2PC$, in the cyclohexane/water system. The lowest interfacial tension values were 2.8, 3 and 12 mN m^{-1}, respectively, indicating that the increase of –CH$_2$ causes the increase in interfacial tension of lipid molecules. The headgroups between DMPE and DPPC are also different. The ethanolamine and the phosphate group in DMPE can form intermolecular hydrogen bonds, and thus lead to higher adsorption energies and lower adsorption activity.

The acquisition of equilibrium interfacial tensions is very important to plot the adsorption isotherms. Extrapolation procedures are often used to obtain equilibrium interfacial tensions for low surfactant concentrations when the time for building the adsorption equilibrium is long [48]. Thus, the equilibrium interfacial tension can be determined by the extrapolation through the $\gamma(1/\sqrt{t})$ plot to infinite time based on the so-called long-time approximation for diffusion-controlled adsorption [49]. The resulting extrapolated γ_{ext} values may be used as a good approximation for the real equilibrium values γ_{equ}. This long-time approximation is given by Hansen–Joos [50–53] as follows:

$$\left[\frac{d\gamma}{d(1/\sqrt{t})}\right]_{t\to\infty} = \frac{RT\Gamma^2}{C_0}\sqrt{\frac{\pi}{4D}} \tag{1.1}$$

where $\gamma(t)$ is the dynamic interfacial tension, C_0 is the surfactant bulk concentration, R is the gas constant, T is the absolute temperature, D is the diffusion coefficient, and Γ is the surface concentration. The linear relationship between $\gamma(t)$ and $1/\sqrt{t}$ remains valid only for a finite period of time, indicating the adsorption process is diffusion-controlled. It is thus necessary to fit the entire experimental curve to the diffusion-controlled model when accurately determining the equilibrium interfacial tension. An adsorption isotherm for DPPC or DMPE can be correspondingly plotted from the extrapolated interfacial tension values (Figure 1.4). At a certain defined value, a critical aggregation concentration (CAC) will be able to appear, but any further concentration increase cannot result in interfacial tension variations.

The model derived by Langmuir and Frumkin can be considered as a theory for simulating the experimental data. The Langmuir adsorption isotherm is the most commonly used equation as well as the basis for most of the adsorption kinetics models for surfactant systems [52]. It is given as:

$$\Gamma = \Gamma_\infty \frac{c_0}{a_L + c_0} \tag{1.2}$$

where Γ_∞ is the saturation adsorption and a_L is the Langmuir adsorption constant indicative of the concentration at which half of the interfacial coverage (at $F = \Gamma_\infty/2$) is reached. Through the fitting of the experimental data by this isotherm equation, one can obtain the values for two characteristic parameters Γ_∞ and a_L. The plotted curve for DPPC is in good agreement with the experimental data, but the curve for DMPE is clearly different from the experimental points at low lipid concentrations. This is because the Langmuir isotherm does not take into account the mutual interactions between adsorbed molecules. To introduce such interactions, the Frumkin isotherm is used to plot the experimental data. The Frumkin isotherm is presented by [45, 52]:

$$\gamma = \gamma_0 - RT\Gamma_\infty\left[\ln\left(1 - \frac{\Gamma}{\Gamma_0}\right) + a'\left(\frac{\Gamma}{\Gamma_\infty}\right)\right] \tag{1.3}$$

and:

$$c_0 = a_F \frac{\Gamma}{\Gamma - \Gamma_\infty} \exp\left(-2a'\frac{\Gamma}{\Gamma_\infty}\right) \tag{1.4}$$

where a_F is the Frumkin adsorption constant, a' is the intermolecular interaction parameter, and γ_0 is the interfacial tension of the pure solvent system (here chloroform/water). For DPPC, the best fitting by the Frumkin isotherm is almost consistent with the result obtained by the Langmuir isotherm (Figure 1.4a). For DMPE, one can also obtain a much ideal plotted result via the Frumkin isotherm.

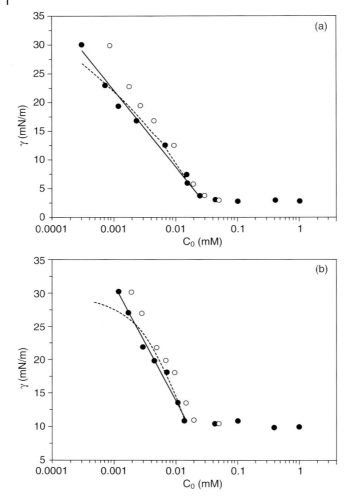

Figure 1.4 The γ–log C_0 curves of phospholipid at the interface of chloroform/water. Open circles, experimental data without concentration correction; filled circles, experimental data with concentration correction; broken line, curve fitted by the Langmuir isotherm; solid line, curve fitted by the Frumkin isotherm. (a) DPPC; (b) DMPE. (Reprinted with permission from [45]. © 1996, Elsevier.)

In this case, hence, the lateral interaction is an important factor to describe the surface state (Figure 1.4b). According to the fundamental adsorption equation of Gibbs [52], the minimum area per molecule can be calculated by getting the slope of the plot γ versus $\ln c$ at the CAC. By such a method, the obtained values for DPPC and DMPE are 61 and 50 Å^2 per molecule, respectively [54]. These values are identical to those obtained by the respective adsorption isotherm.

1.2.4
Protein Layers at the Oil/Water Interface

Proteins, as one of the components of the biological membrane, play key roles in the transportation of substances and the exchange of energy for cells. This is attracting increasing interest for understanding their kinetic and thermodynamic properties at the interface, and further gaining insight into the function and interfacial behavior of proteins. The adsorption kinetics, mainly studied by dynamic surface tension measurements, show many features that deviate from those of typical surfactants caused by the protein structure and chemistry that allow these molecules to change conformations in the adsorbed state [55–60].

1.2.4.1 Kinetics of Protein Adsorption

Several known proteins such as human serum albumin (HSA), β-lactoglobulin (β-LG), and β-casein (β-CA) are selected as model proteins to study their dynamic interfacial tension γ as a function of time t by the pendent drop technique at the chloroform/water interface [61, 62]. The adsorption of proteins at the interface depends significantly on their concentration in the water phase. With increasing protein concentration the interfacial tensions tend to a specific minimum value for each protein, which basically reflects the interfacial activity of the proteins [63]. The time to reach adsorption equilibrium becomes shorter at higher concentrations. However, the rapid decrease of interfacial tension causes a larger area change of the pendent drop, even if the drop volume has been kept unchanged. This implies that the interfacial area cannot be controlled at such conditions. It thus is difficult to explain these data by using a usual diffusion-controlled model.

For lower protein concentrations, the adsorption kinetics can be described by the long-time approximation for diffusion-controlled adsorption [64–67]. In our cases, the Hansen–Joos equation is adopted and the equilibrium or quasi-equilibrium interfacial tension for each protein system can be extrapolated via:

$$\left[\frac{\partial \gamma}{\partial(1/\sqrt{t})}\right]_{t \to \infty} = RT\Gamma^2 \frac{\sqrt{\pi/4D_{\text{eff}}}}{c_0} \quad (1.5)$$

where γ is the dynamic interfacial tension, c_0 is the protein bulk concentration, R is the gas constant, T is the absolute temperature, and Γ is the interfacial excess concentration. D_{eff} is the effective diffusion coefficient that stands not only for the transport of molecules of different bulk conformation to the surface, but also includes possible conformational changes at the interface related to the interfacial tension. The quasi-equilibrium interfacial tension values, $\gamma(t \to \infty)$ of the three proteins can be estimated at each concentration by using Equation 1.5. With increasing the concentration of individual systems, the interfacial tension value decreases gradually until a constant value is reached, which should reflect a state of monolayer coverage of the adsorption layer: γ_{eq} HSA $= 10.5\,\text{mN m}^{-1}$, γ_{eq} β-LG $= 8.2\,\text{mN m}^{-1}$, and γ_{eq} β-CA $= 2.4\,\text{mN m}^{-1}$, respectively. The minimum equilibrium interfacial tension directly corresponds to the interfacial activity of the protein. The smaller this γ_{eq} value, the higher the interfacial activity of the adsorbed molecules. It should be noted that the surface activity can be defined in various ways and here we discuss only the maximum surface pressure $\pi = \gamma_0 - \gamma$ as one of the possible measurements.

The surface pressure isotherms obtained at the oil/water interface are higher than that at the air/water interface. This is attributed to the penetration of protein loops and/or tails into the oil phase, which causes a larger adsorption layer thickness at the oil/water interface than that formed at the surface of an aqueous solution [68–72]. Additionally, by comparing the interfacial tension isotherms with dynamic surface pressure curves for standard proteins like HSA, β-CA and β-LG at the oil/water and air/water interface, it can be deduced that the dynamics of the adsorption process and the equilibrium adsorption characteristics are largely influenced by the nature of the interface. For HSA and β-CA, both the rate of interfacial tension increase and the equilibrium values of π at the oil/water interface are higher than the corresponding values at the air/water interface [73–79]. Comparable values can also be observed for β-LG, but at the low bulk concentration, the increase in π is more distinct at the air/water interface, which is definitely relevant to the resulting conformation of the protein layer.

1.2.4.2 Formation of "Skin-Like" Protein Films on a Curved Interface

Proteins such as HSA, β-LG, and β-CA can form "skin-like" films at a drop surface [61, 80]. The drop is produced by immersing pure chloroform solvent into protein buffer solution through a quartz capillary. After a certain period of time, a folded drop surface can be formed by the adsorption of the protein and the "skin-like" morphology becomes visible upon shrinking the volume of the chloroform drop in the aqueous solution. The formation process is shown in Figure 1.5. The droplet volume decreases gradually with the elapsed adsorption time at the chloroform/

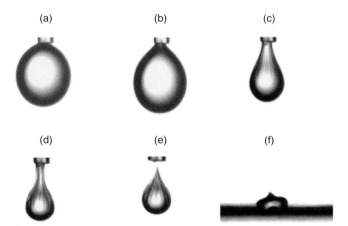

Figure 1.5 Images of droplets covered by β-LG at the chloroform/water interface after different adsorption times: (a) 30, (b) 900, (c) 1340, (d) 1380, (e) 1560, and (f) 1620 s. (Reprinted with permission from [61]. © 2003, Elsevier.)

water interface. Then the folded droplet becomes smaller and forms the skin-like film. With increasing time of adsorption, a neck appears close to the end of the capillary and the drop has a size of hundreds of micrometers. Finally, it drops off to the bottom of the cuvette filled with protein buffer solution. The three example proteins show the same interfacial phenomenon of a folded drop surface. However, the shrinking time for the presence of the skin-like state is totally different. Additionally, the collapse of the folded droplet for HSA is more sphere-like as compared with β-LG and β-CA because HSA keeps a more stable configuration at the curved chloroform/water interface. β-CA needs a longer time to form the skin film, but it quickly shrinks and finally almost disappears.

The dominant thermodynamic driving force for protein adsorption at the drop surface is the removal of nonpolar amino acid side-chains from the aqueous solution [81]. In the adsorbed layers, some unfolded protein molecules can cross-link each other via the strong covalent interaction and such protein adsorption is irreversible [82–84]. HSA and β-LG, which can form stable "skin-like" films, commonly have a globular protein conformation in the native state. The adsorption at the chloroform/water interface may induce an unfolding of the protein, thus leading to an exposure of the hydrophobic residues and making the unfolded protein substantially more hydrophobic than the native one. Moreover, the formation of a globular protein structure is weakened due to the loss of entropy upon protein unfolding [85]. That is, the unfolding of a protein at the interface is an entropy-favorable process and may be the main driving force for the adsorption of proteins at interfaces [86, 87]. Hence, the adsorbed β-LG or HSA exist at the interface in the unfolding conformation and form the stable protein film with covalent cross-links. However, β-CA has a flexible linear structure without intramolecular cross-links [88]. Although it can form a "skin-like" film at the chloroform/water interface, high mechanical resistance is absent in such a film. Dickinson et al. [89] have demonstrated that the shear viscosities of disordered β-CA are two or three orders of magnitude less than those for the globular proteins. If a protein has a more disorder structure and bigger size, its interfacial adsorption rate tends to be reduced. Thus, the interfacial instability of β-CA is likely ascribed to this factor.

1.2.5
Interfacial Behavior of Phospholipid/Protein Composite Layers

The composite structure of phospholipids and proteins is ubiquitous in nature, such as in biological membranes. The interaction between them commonly takes place at the interface. Thus, fabrication of a simple phospholipid/protein composite layer as a biomimetic membrane is of particular importance to obtain information on interfacial behavior of biological membranes. Such a model system can be easily created by using the pendent drop technique (e.g., the adsorption of the two components from different sides of liquid/liquid interface can happen simultaneously when a solution of phospholipid in chloroform is immersed into an aqueous protein solution). The change of the hydrophobicity or hydrophilicity and possible configuration of proteins as it adsorbs from bulk to the interface affects the adsorption activity [90–92]. The interaction between proteins and phospholipids can be studied by interfacial tension measurements at the oil/water interface via the adsorption of the single components from separate bulk phases [80].

1.2.5.1 Dynamic Adsorption and Mechanism

The dynamic adsorption behavior of phospholipids and proteins at the oil/water interface can be measured through the interfacial tension γ as a function of time t. Next, the complex system of DPPC/β-LG is used as an example to illustrate this adsorption process. The progress of the $\gamma(t)$ dependencies is rather different when the protein concentration is varied at constant DPPC content [62, 63, 80]. At lower protein concentrations ($C < 0.1 \,\text{mg}\,\text{l}^{-1}$), the interfacial tension of the complex interfacial layers decreases with increasing DPPC concentration and γ displays a great time dependence. In this case, both DPPC and protein determine the adsorption at the interface because the so-called induction period also appears in the DPPC concentration of $5 \times 10^{-6} \,\text{mol}\,\text{l}^{-1}$ – a condition where the protein controls the adsorption process. As the protein concentration reaches a value of $0.8 \,\text{mg}\,\text{l}^{-1}$, the effect of DPPC on adsorption is smaller. At this condition the protein principally dominates the competitive adsorption of the DPPC/β-LG mixture at the chloroform/water interface. The plateau, which can often be observed in the isotherm of a pure DPPC monolayer [23, 93], vanishes and γ sharply decreases even when the DPPC concentration is rather low. The increase of the DPPC concentration leads to a decrease of the adsorption time that is necessary to reach a constant interfacial tension. The interaction between both may reduce the adsorption energy and allow more molecules to adsorb at the interface. As the DPPC concentration is above the CAC at the constant protein concentration, the adsorption can reach the equilibrium rapidly. As a consequence, the DPPC controls the complete adsorption process and the protein is basically detached from the interface owing to the presence of the much more high interfacially active lipid.

The dipalmitoylphosphatidylethanolamine (DPPE)/β-LG complex system is used as a model to investigate the effects of pH value and ionic strength on the adsorption kinetics of phospholipid/protein at the interface [63, 80]. The dynamic interfacial tension of the mixed DPPE/β-LG layers is measured as a function of time at a constant phospholipid concentration. The results indicate that the increase of ionic strength of the aqueous solution can enhance the adsorption rate of the protein. However, the higher pH value can lead to a reduction of the adsorbed amount. The effect of salt concentration on the adsorption becomes negligible and on this occasion the protein dominantly controls the adsorption when the protein concentration reaches a relatively higher value. The incorporation of protein into the lipid layer involves two steps. The first is the arrival of the protein molecules at the subsurface in which the arrival rate is controlled by the bulk concentration and the diffusion coefficient. The other is the adsorption driven by the interaction with lipid at the interface. In acid aqueous solution of pH 5, the adsorption rate of protein is lower than that in a basic subphase, pH 8. At a higher pH value, which corresponds to the increase of the adsorption rate, the hydrophobic interaction between both promotes the incorporation of proteins into the adsorbed lipid layer. It should be noted that the equilibrium interfacial tension is independent of the pH value. Through the systematic investigation for the salt concentration and pH value effects, it can be found that the electrostatic interaction along with hydrophobic interaction between protein and phospholipid controls the formation of stable complex DPPE/β-LG interfacial layers.

In addition, the competitive adsorption dynamics of the phospholipid/protein complex system at the interface has been investigated by the drop volume method [62]. Protein concentration, conformation, and phospholipid structure are changed to obtain information on influencing the interfacial adsorption behavior. The proteins selected include the HSA, β-LG and β-CA, which possess different conformations. The lipids contain the zwitterionic DPPC, DPPE, dimyristoylphosphatidylcholine (DMPC), and DMPE with different chain lengths or uncharged headgroups. At constant protein concentration, the equilibrium interfacial tension decreases with the increase of lipid concentration. When the lipid concentration is close to CAC, proteins do not produce the pronounced effects on the equilibrium interfacial tension of the complex layers. On the contrary, the structure of the phospholipid can trigger an obvious influence on the equilibrium interfacial tension and its headgroup plays a more important role than the hydrophobic chain.

1.2.5.2 Assembly of "Skin-Like" Complex Films on a Curved Interface

In the above description, it was introduced that the individual protein can form a "skin-like" film at the chloroform/water interface. Actually, the phospholipid/protein complex system also has a similar propensity to form a "skin-like" film that is constructed at a curved drop surface by the coadsorption of two components from different subphases [80, 94]. The time to onset of shrinkage depends on the given bulk concentration. The higher the protein concentration, the faster the time to start shrinking. The addition of phospholipid such as DPPC into chloroform can accelerate this process. At the folding stage of the skin, the formation of a multilayer occurs, which was proved by atomic force microscopy (AFM) by transferring a skin-covered drop to a mica substrate, and then removing the solvents outside and inside by evaporation [80]. With the increasing bulk concentration, the adsorption of pure DPPC at the same interface can also cause a change of the drop shape [45]. However, the "skin-like" structure is never observed even for very high bulk concentrations.

The headgroup of the phospholipid and the pH value has a slight effect on the formation of "skin-like" complex films. The isoelectric point of β-LG is at pH 5.2 [95]. This means that the net negative charges in the protein are enriched in the range of pH > 5.2. Changes in the pH value of the aqueous phase can affect the charged state of β-LG, and thus may lead to a change of the interaction between phospholipid and protein. When a zwitterionic lipid, DPPE, DPPC or the negatively charged phospholipid, dipalmitoylphosphatidic acid (DPPA), is dissolved in the chloroform to coadsorb with β-LG at the interface, the coadsorption rate can be reflected by the adsorption kinetic measurements [94]. The obtained results indicate that the adsorption rate of β-LG at the interface indeed becomes faster in the presence of lipids and the DPPA or DPPE/β-LG systems have a relatively faster adsorption process compared with that of the DPPC/β-LG system. Both lipids, DPPE and DPPA, have smaller headgroups and can be charged with the variation of pH value. Furthermore, β-LG itself possesses both negative and positive charges in a molecular chain. Thus, the encounter of lipid and protein at the interface may neutralize some charges of proteins. As a consequence, slight electrostatic interaction between lipids and proteins may occur at the interface and leads to a quicker adsorption rate. However, the measurement of dynamic surface tensions of DPPC/β-LG shows no noticeable influence on the coadsorption behavior between pH 4 and 6. At pH values between 7 and 8 both the lipid and β-LG carry a negative net charge, and the adsorption rates are similar. This means that the coadsorption is predominantly controlled by the kinetics rather than by electrostatic interactions between the two components.

1.3
Modeling Membrane Hydrolysis *In Vitro*

Hydrolysis of phospholipids is an important physiological process including biosynthesis and degradation of cell membranes. The products generated through hydrolysis like phosphatidic or arachidonic acid play crucial roles in cellular signal transduction [96]. These reactions are catalyzed by different phospholipases, which can be divided into PLA_1, PLA_2, PLC, and PLD according to the site of cleavage on the lipid molecule (Figure 1.6). It is well known that most biological reactions in cells are essentially heterogeneous enzyme catalysis at the membrane–water phase boundary. However, many aspects of the membrane hydrolysis mechanism at interfaces are still unclear. Thus, it is necessary to set up a simple and efficient model membrane to obtain relevant information about membrane hydrolysis. A

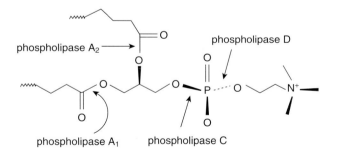

Figure 1.6 Representation of different sites of the phospholipid molecule attacked by phospholipases. PLA_1 and PLA_2 hydrolyze the carboxylic ester linkages in the *sn*-1 and *sn*-2 position of phospholipids. PLC and PLD hydrolyze either of the phosphoric ester bonds of phospholipids to cleave off part of the headgroup. Arrows indicate the sites attacked by the corresponding enzymes.

lipid monolayer at the air/water interface represents a useful model for studying the mechanism of interfacial enzyme reactions.

Such a lipid monolayer at the air/water interface possesses some merits over other lipid model systems. For example, with a spread lipid monolayer at the air/water interface, thermodynamic properties of the interface such as the surface concentrations of the substrate and other constituents and the interfacial tension can be tuned at will by compressing or expanding the monolayer [97]. Thus, this monolayer system enables us to study the influence of these variables on the interfacial recognition and kinetic process of enzymes. A large number of modern microscopic and spectroscopic techniques have been developed to investigate the interfacial reactions of phospholipases at the air/water interface [98]. In this section, we will briefly introduce several experimental techniques that have been adopted widely to study the hydrolysis of lipid monolayers. These methods can provide information on the phase changes of lipid monolayers and hydrolysis processes catalyzed by enzymes.

1.3.1
PLA2

PLA_2 is a type of calcium-dependent enzyme, abundant in living organisms, that hydrolyzes the naturally L-enantiomer of phospholipids within membranes to produce a lysophospholipid and a fatty acid [99]. These enzymes carry out a major role in venoms and digestive fluids, and are also relevant to the regulation of inflammation and the immune response [100–103]. In aqueous solution, PLA_2 has an α-helix-enriched conformation and during the enzyme catalytic reaction a conformation change of PLA_2 may take place as it adsorbs to the lipid monolayer at interface.

From the π–A curves of a lipid monolayer (e.g., L-DPPC), the initial surface pressure changes can lead to phase variations of the monolayer by both molecular rearrangement and PLA_2 adsorption. In order to study the dynamic interaction of lipids/enzymes, surface pressure–time (π–t) curves are often used to follow the whole reactive process "*in situ.*" Surface pressure (π) as a function of time (t) reflects both the adsorption kinetics of PLA_2 as well as L-DPPC monolayer hydrolysis. The π–t curves show that the change of surface pressure is the common result of the adsorption of PLA_2 into the L-DPPC monolayer and the monolayer hydrolysis [104]. One of the hydrolysis products, lysolipid, is water-soluble and can detach from the interface into the subphase, and thus lead to the decrease of surface pressure. Furthermore, the loss of constituents at the interface enables the penetration of the enzyme and results in the increase of surface pressure. The interplay of these factors causes the discernable variation of surface pressure. The catalytic hydrolysis and concomitant adsorption of PLA_2 at the interface can explain the low value of the maximum surface pressure at lower initiating pressure [105]. This also allows us to evidence that the most effective enzyme activity of PLA_2 is in the liquid-expanded/liquid-condensed coexistence region.

Polarization-modulated external IR reflection absorption (PM-IRRAS) spectroscopy [98] has been proved to be a versatile approach for studying the hydrolysis reaction of lipids at the air/water interface. Through PM-IRRAS measurement of an l-DPPC monolayer we can analyze the reactants and products at the interface [106, 107]. The intensity of some characteristic adsorptions such as the C=O stretching vibration, CH_2-scissoring mode, and hydrated and nonhydrated phosphate groups considerably depend on the lateral density of the molecules and the orientation of the transition moments in the plane of incidence. It is thus possible to quantitatively monitor the hydrolysis reaction by the measurement of intensity variations of the adsorption bands [108, 109]. In addition, synchrotron grazing incidence X-ray diffraction (GIXRD) is also applicable for determining the structure changes of a lipid monolayer with or without the occurrence of enzyme [105, 110–112]. For instance, the monolayer structure of l-DPPC was greatly affected by PLA_2 adsorption. Compared with the pure l-DPPC monolayer, the tilted angle is much smaller when PLA_2 adsorbs at the interface. The GIXRD measurement demonstrates that the headgroup properties such as orientation and hydration, which determine molecular area in the condensed DPPC monolayers, change due to the specific interaction between the lipid and enzyme. The electrostatic force between enzyme molecules and the domain boundary can also affect the orientation of the lipid molecules and further change their conformation. Obviously, the higher hydrolysis efficiency correlates with a larger structural change at the interface upon the PLA_2 adsorption.

Brewster angle microscopy (BAM) is a powerful tool to follow the morphology changes of phospholipid domains in a monolayer during the hydrolysis catalyzed by enzyme [104, 113]. This technique allows for characterization of the long-range orientation order of lipid monolayers that arises from the optical anisotropy induced by the tilted aliphatic chains [114, 115]. Light cannot be reflected from the interface because of their different refractive indices as *p*-polarized light reaches the air/water interface at the Brewster angle. Under such a constant angle of incidence, a lipid monolayer can alter the Brewster angle conditions and light will be reflected. It can thus be utilized to record and image. The different shapes in the domain observed by the BAM measurement are relevant to different phases of the monolayer such as liquid-expanded, liquid-condensed, and solid phases. The formation of different morphological domains correlates with the increase of anisotropy. This means that the cohesion of molecules into a domain is favorable in a defined direction. Usually, l-DPPC is used as a model compound for studying the hydrolysis of the monolayer membrane because it is a principle component of cell membranes. BAM images of l-DPPC monolayers attacked by PLA_2 are shown in Figure 1.7. The hydrolysis reaction starts either in the domain or at the edge in dependence on the preparation conditions of the lipid monolayer. After the injection of PLA_2, the domains damaged in the inner part can be observed clearly. One of the reaction products, lysolipid, is water-soluble and leaves from the interface. This thus makes more free space available. As a consequence, the ratio of the ordered region becomes smaller, which is reflected in the change of domains.

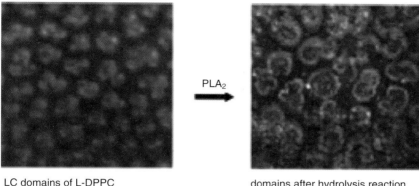

| LC domains of L-DPPC | domains after hydrolysis reaction |

Figure 1.7 BAM images of a L-DPPC monolayer at the air/water interface before the injection of PLA$_2$ into the subphase and after the hydrolysis reaction. (Reprinted with permission from [104]. © 2000, Wiley.)

From the above multiple characterized results, we can determine the avenue to hydrolyze the lipid membrane by PLA$_2$. The first step is the adsorption of enzyme at the lipid monolayer as a molecular recognition process and then the hydrolysis reaction occurs by cleaving the alky chains of the lipid. Furthermore, the PLA$_2$ can preferentially hydrolyze the condensed domains with the same molecular chain orientation.

Additionally, the influence of indole inhibitors on the hydrolysis reaction of L-DPPC monolayers catalyzed by PLA$_2$ was also explored by using the π–t curves and BAM measurement [116, 117]. The addition of a PLA$_2$ inhibitor to the lipid monolayer leads to a fast increase of the surface pressure after enzyme injection. However, the surface pressure value remains unchanged for a long time after it reaches the maximum. During this period, the domain shape and number density of the lipid monolayer are not varied, indicating the inhibitor of PLA$_2$ takes effect and this process can persist for a certain period of time. Inhibitors synthesized for inhibiting PLA$_2$ activity in the enzymatic reaction hold potential for developing new drugs in the treatment of diseases.

1.3.2
PLC

PLC is a water-soluble enzyme. It can catalyze the hydrolysis of the ester bond in the C-3 position of a glycerophospholipid, such as L-DPPC, and produce a water-insoluble dipalmitoylglycerol (DPG) and a water-soluble phosphocholine [118]. Unlike PLA$_2$, which is able to cause a rapid change of membrane composition in the presence of Ca^{2+}, the initial hydrolysis rate of PLC to lipids is remarkably lower for a certain lag time. After this period, the enzymatic activity has an abrupt increase. This behavior is called lag–burst enzymatic cleavage [119, 120].

As a direct optical visualization means, BAM can be used to trace this hydrolysis process, finding that PLC already starts to catalyze the hydrolysis reaction of the lipid monolayer at the beginning of the lag period [121]. During the hydrolysis process, "lotus-like" domains with different sizes can be observed. Phosphocholine, as one of the hydrolysis products, is water-soluble and can enter the subphase after the enzymatic reaction. After L-DPPC is hydrolyzed into DPG, the tilt angle of the aliphatic chains tends to zero while the resulting molecules are compressed at the interface [122]. Hence, the morphological change of the lipid monolayer is mainly provoked by the generated DPG and the nonreacted L-DPPC at the interface. The dissolution of phosphocholine in the bulk phase and molecular rearrangements of L-DPPC and DPG at the interface will lead to a reduction of molecule density of condensed domains and an increase of surface pressure. BAM images also confirm that lipid monolayer hydrolysis by PLC preferentially occurs in the condensed phase.

In addition, the PM-IRRAS spectra allow us to obtain information on the molecular vibrations at the monolayer interface. The changes of P=O stretching band intensity can be used to quantitatively monitor the hydrolysis process of lipid monolayers catalyzed by PLC and assess the enzyme activity [123, 124].

1.3.3
PLD

PLD is an abundant enzyme that catalyzes the cleavage of the terminal phosphate ester bond at the polar headgroup of phosphatidylcholine to phosphatidic acid (PA) and a water-soluble choline. This enzyme is found in plants, bacteria, and animals, and is involved in a variety of cellular processes such as lipid metabolism, vesicle trafficking, and signal transduction [125, 126]. Furthermore, PA is an important source of arachidonic acid for the synthesis of prostaglandins and leukotrienes.

The π–t isotherms show that the variation of surface pressure during the hydrolysis reaction catalyzed by PLD is relatively small compared to those by PLA_2 and PLC [127–129]. PM-IRRAS and BAM techniques can also be used to follow the hydrolysis process. Due to the diversity of PM-IRRAS spectra between DPPA and L-DPPC, it is easy to determine the components at the lipid monolayers by the PM-IRRAS measurement. The results confirm that the intensities of υ_s (P=O) and υ_{as} (P=O) vibrations linearly depend on the mole fraction of L-DPPC. Such linearity is often expected for ideal mixtures or phase-separated systems.

From BAM images, it can be observed that some dark dots appear in the inner liquid-condensed phase domains and there is a "lotus" or "Swiss cheese" structure that forms gradually with an increase of reaction time [128]. The formation of the "lotus" domains indicates the appearance of a new phase in the DPPC condensed domain [130]. Actually, DPPA plays a key role in the formation of the new phase because the other reaction product is water-soluble. This is further demonstrated by the combination of π–A isotherms and BAM images. The mixtures of L-DPPC and DPPA at different molar ratios have an obvious phase separation [130, 131]. As the surface pressure is below the plateau in the mixtures with few DPPA molecules, small domains can be observed by BAM, indicating a phase separation in the lipid monolayer in which DPPA-rich domains are surrounded by disordered L-DPPC. In the lag phase the percentage of reaction product DPPA is very small but slowly accumulates owing to phase separation. Production of DPPA accelerates the adsorption of PLD at the lipid interface by electrostatic attraction and thus initiates the fast hydrolysis of L-DPPC.

The activity of PLD strongly depends on the presence of Ca^{2+}, which has been related to promoting PLD binding to lipid interfaces and enhancing the catalytic reaction [132, 133]. As illustrated above, as one of the reaction products, DPPA, is the lipidic activator of PLD. It has been suggested that a lateral separation of PA induced by Ca^{2+} can activate the enzyme and lead to the lag–burst behavior of PLD [134, 135]. Recently, by using various lipid monolayers, Brezesinski et al. proved that the phase separation of PA is the essential requirement for the production of high PLD activity [136]. The corresponding mechanism is shown in Figure 1.8(a). Once a critical amount of PC is transformed into PA by the fundamental enzyme activity during the lag phase, the hydrolysis product becomes incompatible with the original lipid monolayer and segregates into PA-enriched domains. PLD can interact with PA-rich domains with the help of Ca^{2+} and a burst of enzymatic activity occurs. Therefore, PA segregation with lipid layers leads to the activation of PLD. In addition, the hydrolysis reaction can also be inhibited by phase segregation. In this case, the PA-rich domains adopt an upright alignment of lipid chains (Figure 1.8b). The changed orientation of the PA headgroup can affect the adsorption of enzyme and hinder PLD to further cleave the PC neighboring the PA-rich domains. Up to now, all measurement results show that PLD maximum activity exists in a more disordered phase, indicating that fluidity and defects in the monolayer are more important than a preorientation of the substrate induced by enzyme adsorption, like PLA_2.

All in all, from the above description of hydrolysis reactions catalyzed by various enzymes including PLA_2, PLC as well as PLD, it can be concluded that the lipid monolayer is a good model to help us to understand the phospholipid membrane hydrolysis process and mechanism at the interface. The multiple techniques and methods developed, such as BAM, PM-IRRAS, GIXRD, and so on, can be used to trace the enzymatic reactions at lipid membranes and provide relevant information on the interfacial interactions during the reaction.

1.4
Polyelectrolyte-Supported Lipid Bilayers

Phospholipid membranes, as an important component of cell structures, can naturally create a boundary to separate the internal environment from the external environment of cells. To better understand the structure, function, and properties of biological membranes, researchers have attempted to develop various biomimetic models, well known as lipid vesicles (liposomes). Due to the amphiphilic nature of lipid molecules, they can spontaneously form hollow spherical closed

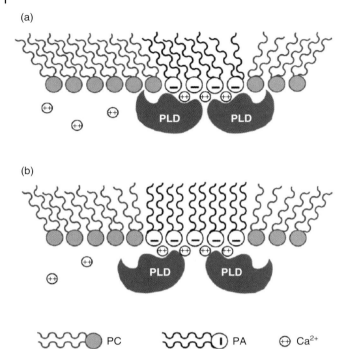

Figure 1.8 Model of the regulation of PLD. (A) Activation of PLD through domain formation of PA. Once a critical amount of PC is converted into PA by the basal activity of the enzyme during the lag phase, the product becomes immiscible with the substrate and segregates into PA-rich domains. In the presence of calcium ions, PLD can interact with PA-rich domains and a burst of the enzymatic activity occurs. Hence, product segregation within lipid membranes activates PLD. (B) Inhibition of PLD through PA-rich domains adopting an upright alignment of the lipid chains. This phase state indicates a changed orientation of the PA headgroup that complementarily affects the orientation of bound enzyme and impedes PLD to continue the hydrolysis of PC neighboring the PA-rich domains. (Reprinted with permission from [4]. © 2008, Elsevier.)

bilayers in water. Up to now, such biomimetic vesicles have not only served as model biological membranes in our outstanding of basic cell physiology, but have also been expanded to numerous applications such as drug vehicles, gene therapy, and nanoreactors [137–140].

Recently, supported lipid bilayers have been developed as a new type of biomimetic model system because of the emergence of a number of surface-sensitive characterization techniques [141–144]. The solid substrates that are usually used to deposit or confine lipid bilayers include inorganic surfaces (e.g., glass, silica, micas, metals, etc.), self-assembled alkanethiol monolayers, polymers, and polyelectrolytes. Methods to create such biomimetic systems involve the LB technique

[145–148] and self-assembly technique through lipid vesicle fusion [149, 150]. In the latter case, lipids are adsorbed onto the substrates in the mode of vesicles before they spread over the surfaces. The features can be characterized by single particle light scattering and energy transfer measurements [151–154]. There have been plenty of reviews to summarize lipid membranes supported on inorganic surfaces [155, 156], self-assembled alkanethiol monolayers [157], and natural polymers [6, 158]. Therefore, here, we will mainly concentrate on the development and application of lipid bilayers supported by polyelectrolyte multilayers fabricated from the LbL assembly technique.

1.4.1
Polyelectrolyte Multilayers on Planar Surfaces

Polyelectrolyte multilayers offer a highly versatile surface coating with adjustable properties depending on the components used and build-up conditions during the LbL assembly. The use of polyelectrolyte multilayers provides a distinct advantage over that of other solid substrates. According to specific requirements, different functional polyelectrolytes can be selected and directly deposited from solution to a variety of substrates by LbL adsorption. The chemical and physical properties of polyelectrolyte multilayers are well defined. Furthermore, this method allows for remarkable control over the resulting polyelectrolyte multilayer thickness, even down to molecular dimensions, which can be realized by varying the number of adsorption cycles, ion strength, and pH [159]. More intriguing, the LbL method can also be utilized to form hollow polyelectrolyte microcapsules by alternatively coating the opposite components on the removable colloidal particles. The capsule coated with functional lipid bilayers is analogous to an "artificial cell," which has been considered for wide-range applications in biomimetic delivery vessels, drug-controlled release, biosensors, and biodevices. The greater promise is aimed at developing an intelligent biomimetic sac system for targeting drug delivery and consideration as a model cell for investigating membrane-related biological functions.

The concept of a polyelectrolyte-supported lipid membrane, expected to act as a model system for studying the complex properties of biological membrane, was first put forward by Möhwald *et al.* [160]. The lipid bilayers with 10% charged dioleoylphosphatidic acid and zwitterionic DMPC were fabricated on the polyelectrolyte multilayers by vesicle fusion into bilayers and deposition from monolayers by the Langmuir–Schafer technique (Figure 1.9). Such supported lipid bilayers displayed inherent properties of the biological membrane such as homogeneity and mobility. Subsequently, Knoll *et al.* [161] also confirmed the formation of lipid bilayers on a polyelectrolyte cushion by multiple characterization techniques including time-dependent surface plasmon spectroscopy [162], neutron reflectometry [163], impedance spectroscopy [164], as well as fluorescence recovery after photobleaching (FRAP) [165]. Recently, Delajon *et al.* [166] showed that it was feasible to fabricate a lipid bilayer from DMPC on planar poly(allylamine hydrochloride) (PAH)/poly(sodium 4-styrenesulfonate) (PSS) polyelectrolyte

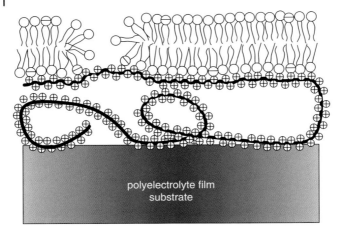

Figure 1.9 Schematic picture of a bilayer attached to a polyelectrolyte support. Lateral fluid-like transport is possible for a sufficiently small fraction of negatively charged anchoring lipids. The polymer appears to create conducting defects. (Reprinted with permission from [160]. © 1999, Elsevier.)

multilayers. The authors adopted the neutron reflectometry technique, which can provide detailed information on the structure and composition of the adsorbed layers, to study adsorption at surfaces, and found that DMPC bilayers could only be favorable at the negatively charged PSS-terminated surface and the following polyelectrolyte layer deposition may also be achieved, but a negatively charged PSS layer has to be first assembled. Although the mechanism of the formation of lipid bilayers on the polyelectrolyte multilayers is not completely understood yet, the current experimental results confirm the feasibility for forming biologically active lipid membranes on such substrates. The soft polyelectrolyte cushion provides a hydrated space for the membrane proteins incorporated on the planar surfaces. Compared with the solid-supported membranes, it can avoid direct interactions between membrane proteins and solid surfaces, which likely cause the risk of protein denaturation [149, 167, 168]. The hydrated polyelectrolyte cushion that resides between lipid bilayers and solid surfaces helps shield the protein with the substrate and protect peripheral sections of the protein. Hence, polyelectrolyte-supported lipid bilayers are emerging as a popular choice for model membranes.

1.4.2
Polyelectrolyte Multilayers on Curved Surfaces

LbL-assembled polyelectrolyte multilayer-supported lipid membranes are not merely limited to planar substrates. Curved surfaces such as colloidal particles coated with polyelectrolyte can also be selected as the support for lipid bilayers [169]. This provides a significant advantage over the use of planar surfaces because some particle-based experimental techniques, such as fluorescent-activated cell sorting, can be adapted to such conditions [170]. The spherical supported lipid bilayer on the polyelectrolyte microcapsule is analogous to the cell membrane. That is to say, to some extent such lipid biointerfacing microcapsules mimic cell membrane structures. Therefore, this type of biomimetic system with a hollow inner chamber cannot only be used as a model to study membrane properties, but may also have some new functions for potential applications in biomedical fields.

Due to the introduction of lipid bilayers on the outer surface of the microcapsule, the original properties of the capsule, such as permeability and biocompatibility, can be altered. For instance, permeability of the capsules before and after coating lipid bilayers is obviously different [171–173]. The following assay was carried out by our group. A fluorescent dye, 6-carboxyfluorescein (insoluble in lipid membranes), was selected as the probe for monitoring the permeability of capsules with the help of the confocal scanning laser microscopy (CLSM) technique [171]. CLSM observation provides evidence of the decrease of capsule permeability after coating the lipid bilayers, which can remain impermeable for at least 6 months (at 4 °C). By comparison with the capsules without coating lipid bilayers, the confocal results indicated that the continuous lipid shell restrained the permeation of the fluorescent probe into capsules. Furthermore, the permeability of such lipid bilayer-coated capsules can be adjusted easily by introducing the enzyme catalytic reaction [174]. This strategy is simple and is easy to perform. An enzyme, PLA_2, which exists extensively in living organisms, was added to the lipid-coated capsule dispersion system. PLA_2 can stereoselectively hydrolyze the *sn*-2 ester linkage of enantiomeric L-phospholipids (e.g., L-DPPC) to generate fatty acids and lysophospholipids. The enzyme reaction results into the rupture of the lipid membrane on the capsules and forms new channels for molecule permeation, thus leading to an improvement of permeability. As a consequence, it is possible to tune the function and property of capsules by the approach of coating lipid bilayers. Such flexible systems can provide promising opportunities for the controlled delivery and release of drugs.

Additionally, the support of lipid bilayers on one-dimensional LbL-assembled polyelectrolyte nanotubes has been achieved [175]. A poly(ethyleneimine) and 3,4,9,10-perylenetetracarboxylicdianhydride $(PEI/PTCDA)_6$ nanotube was assembled by the LbL technique and then the lipid bilayers were formed on the hydrophilic tubular surface by the large unilamellar vesicle fusion method (Figure 1.10a). In this system, the electrostatic attraction between the nanotube support and the lipid membrane is responsible for the formation of wrapping lipid bilayers. Meanwhile, the repulsive force between the liposomes is advantageous to impede the formation of the multibilayer membrane on the surface of the nanotubes. Such supported lipid membranes retain one of the most important features – diffusion along the bilayer plane. This conclusion was demonstrated by the FRAP experiment. Figure 1.10(b) shows the recovery curve of fluorescence intensity with time in the bleached region, indicating a single-exponential relationship between them and more than 80% recovery efficiency. The biointerfacing nanotubes offer a possibility to develop a range of functional bioorganic nanomaterials on the basis of one-dimensional lipid bilayers.

(a)

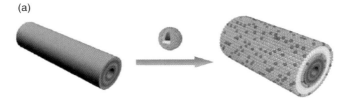

(b)

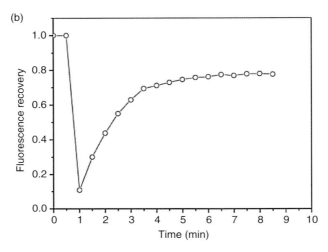

Figure 1.10 (a) Schematic of lipid bilayer coating on LbL-assembled luminescent nanotube by liposome fusion. (b) Fluorescence recovery curves after photobleaching (FRAP) of lipid-coated nanotubes. (Reprinted with permission from [175]. © 2009, Royal Society of Chemistry.)

1.5
Conclusions and Perspectives

This chapter has been written from the perspective of what is a biomimetic membrane, how to fabricate such a biomimetic membrane, and the extent that it can function as a model system for investigating the interfacial behavior and function of biological membranes. More importantly, entities with a lipid bilayer coating will find potential applications in drug delivery and controlled release, biosensors, biocatalysis and bioreactors due to the enhancement of biological availability.

The lipid monolayer at the interface, fabricated by the self-assembly method, can serve as a most simplified model membrane. On the one hand, such a monolayer membrane at the oil/water interface can also be used to investigate the dynamic adsorption and interfacial behavior of phospholipids and phospholipid/protein mixtures. Also, upon the interfacial adsorption of phospholipid and protein molecules, a biomimetic "skin-like" crumpled film structure can be formed. On the other hand, the lipid monolayer spread at the air/water interface is a suitable model membrane that has been applied for the investigation of membrane hydrolysis catalyzed by various enzymes. This simplified model enables us to easily gain insights into chemical reactions involved in enzyme catalysis in the biological membrane. As such, the supported lipid bilayer on the polyelectrolyte multilayers provides a new alternative means to study cellular surface chemistry, including cell signaling, ligand–receptor interactions, enzymatic hydrolysis as well as biosensors *in vitro*. Additionally, biomimetic membrane interfacial nanomaterials will undoubtedly be promising for development and application in the field of both biotechnology and nanotechnology.

In the following chapter, we will illustrate in detail the fabrication, function, and application of biomimetic microcapsules.

References

1. Alberts, B., Johnson, A., Lewis, J., Raff, M., Roberts, K., and Walters, P. (2002) *Molecular Biology of the Cell*, 4th edn, Garland Science, New York.
2. Silvius, J.R. and Nabi, I.R. (2006) Fluorescence-quenching and resonance energy transfer studies of lipid microdomains in model and biological membranes [Review]. *Mol. Membr. Biol.*, **23**, 5–16.
3. He, Q., Zhang, Y., Lu, G., Miller, R., Möhwald, H., and Li, J.B. (2008) Dynamic adsorption and characterization of phospholipid and mixed phospholipid/protein layers at liquid/liquid interfaces. *Adv. Colloid Interface Sci.*, **140**, 67–76.
4. Wagner, K. and Brezesinski, G. (2008) Phospholipases to recognize model membrane structures on a molecular length scale. *Curr. Opin. Colloid Interface Sci.*, **13**, 47–53.
5. He, Q. and Li, J.B. (2007) Hydrolysis characterization of phospholipid monolayers catalyzed by different phospholipases at the air–water interface. *Adv. Colloid Interface Sci.*, **131**, 91–98.
6. Tanaka, M. and Sackmann, E. (2005) Polymer-supported membranes as models of the cell surface. *Nature*, **437**, 656–663.
7. Castellana, E.T. and Cremer, P.S. (2006) Solid supported lipid bilayers: from biophysical studies to sensor design. *Surf. Sci. Rep.*, **61**, 429–444.
8. Yang, T.L., Baryshnikova, O.K., Mao, H.B., Holden, M.A., and Cremer, P.S. (2003) Investigations of bivalent antibody binding on fluid-supported phospholipid membranes: the effect of hapten density. *J. Am. Chem. Soc.*, **125**, 4779–4784.
9. Pum, D., Stangl, G., Sponer, C., Riedling, K., Hudek, P., Fallmann, W., and Sleytr, U.B. (1997) Patterning of monolayers of crystalline S-layer proteins on a silicon surface by deep ultraviolet radiation. *Microelectron. Eng.*, **35**, 297–300.
10. Yang, T.L., Jung, S.Y., Mao, H.B., and Cremer, P.S. (2001) Fabrication of phospholipid bilayer-coated microchannels for on-chip immunoassays. *Anal. Chem.*, **73**, 165–169.
11. Plant, A.L., Brighamburke, M., Petrella, E.C., and Oshannessy, D.J. (1995) Phospholipid alkanethiol bilayers for cell-surface receptor studies by surface-plasmon resonance. *Anal. Biochem.*, **226**, 342–348.
12. Kasahara, K. and Sanai, Y. (2001) Involvement of lipid raft signaling in ganglioside-mediated neural function. *Trends Glycosci. Glycotechnol.*, **13**, 587–594.
13. Stoddart, A., Dykstra, M.L., Brown, B.K., Song, W.X., Pierce, S.K., and

Brodsky, F.M. (2002) Lipid rafts unite signaling cascades with clathrin to regulate BCR internalization. *Immunity*, **17**, 451–462.

14 Qi, S.Y., Groves, J.T., and Chakraborty, A.K. (2001) Synaptic pattern formation during cellular recognition. *Proc. Natl. Acad. Sci. USA*, **98**, 6548–6553.

15 Cornell, B.A., Braach-Maksvytis, V.L.B., King, L.G., Osman, P.D.J., Raguse, B., Wieczorek, L., and Pace, R.J. (1997) A biosensor that uses ion-channel switches. *Nature*, **387**, 580–583.

16 Stenger, D.A., Fare, T.L., Cribbs, D.H., and Rusin, K.M. (1992) Voltage modulation of a gated ion channel admittance in platinum-supported lipid bilayers. *Bioses. Bioelectron.*, **7**, 11–20.

17 Naumann, R., Jonczyk, A., Kopp, R., van Esch, J., Ringsdorf, H., Knoll, W., and Graber, P. (1995) Incorporation of membrane-proteins in solid-supported lipid layers. *Angew. Chem. Int. Ed. Engl.*, **34**, 2056–2058.

18 Möhwald, H. (1990) Phospholipid and phospholipid–protein monolayers at the air/water interface. *Annu. Rev. Phys. Chem.*, **41**, 441–476.

19 Ransac, S., Moreau, H., Riviere, C., and Verger, R. (1991) Monolayer techniques for studying phospholipase kinetics. *Methods Enzymol.*, **197**, 49–65.

20 Zhang, H.J., Wang, X.L., Cui, G.C., and Li, J.B. (2000) Stability investigation of the mixed DPPC/protein monolayer at the air–water interface. *Colloids Surf. A*, **175**, 77–82.

21 Weis, R.M. and McConnell, H.M. (1984) Two-dimensional chiral crystals of phospholipid. *Nature*, **310**, 47–49.

22 Kaganer, V.M., Möhwald, H., and Dutta, P. (1999) Structure and phase transitions in Langmuir monolayers. *Rev. Mod. Phys.*, **71**, 779–819.

23 Kurihara, K. and Katsuragi, Y. (1993) Specific inhibitor for bitter taste. *Nature*, **365**, 213–214.

24 Knobler, C.M. and Schwartz, D.K. (1999) Langmuir and self-assembled monolayers. *Curr. Opin. Colloid Interface Sci.*, **4**, 46–51.

25 McConnell, H.M. (1991) Structures and transitions in lipid monolayers at the air–water interface. *Ann. Rev. Phys. Chem.*, **42**, 171–195.

26 Brezesinski, G. and Möhwald, H. (2003) Langmuir monolayers to study interactions at model membrane surfaces. *Adv. Colloid Interface Sci.*, **100–102**, 563–584.

27 Miller, R. and Kretzschmar, G. (1991) Adsorption kinetics of surfactants at fluid interfaces. *Adv. Colloid Interface Sci.*, **37**, 97–121.

28 Li, J.B., Chen, H., Wu, J., Zhao, J., and Miller, R. (1999) The structure and dynamic properties of mixed adsorption and penetration layers of alpha-dipalmitoylphosphatidylcholine/beta-lactoglobulin at water/fluid interfaces. *Colloids Surf. B*, **15**, 289–295.

29 Zhang, Y., Yan, L.L., Bi, Z.B., and J.B. Li (2001) Dynamic study of the interaction between beta-lactoglobulin and phospholipids during complex film formation. *Acta Chim. Sin.*, **59**, 659–664.

30 Watts, A. (1993) *Protein/Lipid Interactions: New Comprehensive Biochemistry*, Elsevier, Amsterdam.

31 Aksenenko, E.V., Kovalchuk, V.I., Fainerman, V.B., and Miller, R. (2006) Surface dilational rheology of mixed adsorption layers at liquid interfaces. *Adv. Colloid Interface Sci.*, **122**, 57–66.

32 Miller, R., Loglio, G., Tesei, U., and Schano, K.H. (1991) Surface relaxations as a tool for studying dynamic interfacial behavior. *Adv. Colloid Interface Sci.*, **37**, 73–96.

33 Miller, R., Krägel, J., Wüstneck, R., Wilde, P.J., Li, J.B., and Fainerman, V.B. (1998) Adsorption kinetics and rheological properties of food proteins at air/water and oil/water interfaces. *Nahrung*, **42**, 225–228.

34 Miller, R., Li, J.B., Bree, M., Loglio, G., Neumann, A.W., and Möhwald, H. (1998) Interfacial relaxation of phospholipid layers at a liquid–liquid interface. *Thin Solid Films*, **327**, 224–227.

35 Bonfillon, A. and Langevin, D. (1993) Viscoelasticity of monolayers at oil–water interfaces. *Langmuir*, **9**, 2172–2177.

36 Cheng, P., Li, D., Boruvka, L., Rotenberg, Y., and Neumann, A.W. (1990) Automation of axisymmetric drop shape analysis for measurement of interfacial tensions and contact angles. *Colloids Surf.*, **43**, 151–167.

37 Li, J.B., Miller, R., Vollhardt, D., Neumann, A.W., and Möhwald, H. (1995) In *Short and Long Chains at Interfaces* (eds J. Dailant, P. Guenoun, C. Marques, P. Muller, and J. Tran Thanh Van), Frontieres, Paris, p. 201.

38 Li, J.B., Miller, R., Wüstneck, R., Möhwald, H., and Neumann, A.W. (1995) Use of pendent drop technique as a film balance at liquid–liquid interfaces. *Colloids Surf. A*, **96**, 295–299.

39 Zhao, J., Vollhardt, D., Brezesinski, G., Siegel, S., Wu, J., and Li, J.B. (2000) Effect of protein penetration into phospholipid monolayers: morphology and structure. *Colloids Surf. A*, **168**, 287–296.

40 Li, J.B., Kretzschmar, G., Miller, R., and Möhwald, H. (1999) Viscoelasticity of phospholipid layers at different fluid interfaces. *Colloids Surf. A*, **149**, 491–497.

41 Li, J.B., Miller, R., and Möhwald, H. (1996) Phospholipid monolayers and their dynamic interfacial behaviour studied by axisymmetric drop shape analysis. *Thin Solid Films*, **284–285**, 357–360.

42 Kretzschmar, G., Li, J.B., Miller, R., and Motschmann, H. (1996) Characterisation of phospholipid layers at liquid interfaces. 3. Relaxation of spreading phospholipid monolayers under harmonic area changes. *Colloids Surf. A*, **114**, 277–285.

43 Kwok, D.Y., Vollhardt, D., Miller, R., Li, D., and Neumann, A.W. (1994) Axisymmetrical drop shape analysis as a film balance. *Colloids Surf. A*, **88**, 51–58.

44 Li, J.B., Miller, R., Vollhardt, D., Weidemann, G., and Möhwald, H. (1996) Isotherms of phospholipid monolayers measured by a pendant drop technique. *Colloid Polym. Sci.*, **274**, 995–999.

45 Li, J.B., Miller, R., and Möhwald, H. (1996) Characterisation of phospholipid layers at liquid interfaces. 1. Dynamics of adsorption of phospholipids at the chloroform/water interface. *Colloids Surf. A*, **114**, 113–121.

46 Li, J.B., Miller, R., and Möhwald, H. (1996) Characterisation of phospholipid layers at liquid interfaces. 2. Comparison of isotherms of insoluble and soluble films of phospholipids at different fluid/water interfaces. *Colloids Surf. A*, **114**, 123–130.

47 Johnson, M.C.R., and Saunders, L. (1973) Time-dependent interfacial tensions of a series of phospholipids. *Chem. Phys. Lipids*, **10**, 318–327.

48 Fainerman, V.B., Miller, R., and Aksenenko, E.V. (2002) Simple model for prediction of surface tension of mixed surfactant solutions. *Adv. Colloid Interface Sci.*, **96**, 339–359.

49 Fainerman, V.B., Makievski, A.V., and Miller, R. (1993) The measurement of dynamic surface tensions of highly viscous liquids by the maximum bubble pressure method. *Colloids Surf.*, **75**, 229–235.

50 Rillaerts, E. and Joos, P. (1982) Rate of demicellization from the dynamic surface tensions of micellar solutions. *J. Phys. Chem.*, **86**, 3471–3478.

51 Hansen, R.S. (1960) The theory of diffusion controlled absorption kinetics with accompanying evaporation. *J. Phys. Chem.*, **64**, 637–641.

52 Dukhin, S.S., Kretzschmar, G., and Miller, R. (1995) *Dynamics of Adsorption at Liquid Interfaces 1: Theory, Experiment, Application*, Studies in Interface Science, Elsevier, Amsterdam.

53 Miller, R., Sedev, R., Schano, K.H., Ng, C., and Neumann, A.W. (1993) Relaxation of adsorption layers at solution air interfaces using axisymmetrical drop-shape analysis. *Colloids Surf.*, **69**, 209–216.

54 Miller, R., Li, J.B., Wustreck, R., Krägel, J., Clark, D., and Neumann, A.W. (1995) In *Short and Long Chains at Interfaces* (eds J. Dailant, P. Guenoun, C. Marques, P. Muller, and J. Tran Thanh Van), Frontieres, Paris, p. 195.

55 Wang, X.L., Zhang, Y., Wu, J., Wang, M.Q., Cui, G.C., Li, J.B., and Brezesinski, G. (2002) Dynamical and

morphological studies on the adsorption and penetration of human serum albumin into phospholipid monolayers at the air/water interface. *Colloids Surf. B*, **23**, 339–347.
56 Zhang, H.J., Cui, G.C., and Li, J.B. (2002) Morphological investigation of mixed protein/phospholipid monolayers. *Colloids Surf. A*, **201**, 123–129.
57 Ghosh, S. and Bull, H.B. (1963) Surface potentials of protein solutions. *Biochemistry*, **2**, 411.
58 Graham, D.E. and Philips, M.C. (1979) Proteins at liquid interfaces: I. Kinetics of adsorption and surface denaturation. *J. Colloid Interface Sci.*, **70**, 403–414.
59 Kalischewski, K. and Schügerl, K. (1979) Investigation of protein foams obtained by bubbling. *Colloid Polymer Sci.*, **257**, 1099–1110.
60 Krägel, J., Wüstneck, R., Husband, F., Wilde, P.J., Makievski, A.V., and Grigoriev, D.O. (1999) Properties of mixed protein/surfactant adsorption layers. *Colloids Surf. B*, **12**, 399–407.
61 Lu, G., Chen, H., and Li, J.B. (2003) Forming process of folded drop surface covered by human serum albumin, beta-lactoglobulin and beta-casein, respectively, at the chloroform/water interface. *Colloids Surf. A*, **215**, 25–32.
62 Wu, J., Li, J.B., Zhao, J., and Miller, R. (2000) Dynamic characterization of phospholipid/protein competitive adsorption at the aqueous solution/chloroform interface. *Colloids Surf. A*, **175**, 113–120.
63 Yan, L.L., Zhang, Y., Cui, G.C., and Li, J.B. (2000) pH value and ionic strength effects on the adsorption kinetics of protein/phospholipid at the chloroform/water interface. *Colloids Surf. A*, **175**, 61–66.
64 Fleer, G.J. and Scheutjens, H.M. (1982) Adsorption of interacting oligomers and polymers at an interface. *Adv. Colloid Interface Sci.*, **16**, 341–359.
65 MacRitchie, F. (1986) Spread monolayers of proteins. *Adv. Colloid Interface Sci.*, **25**, 341–385.
66 Fainerman, V.B., Lucassen-Reynders, E.H., and Miller, R. (1998) Adsorption of surfactants and proteins at fluid interfaces. *Colloids Surf. A*, **143**, 141–165.
67 Joos, P. (1975) Approach for an equation of state for adsorbed protein surfaces. *Biochim. Biophy. Acta*, **375**, 1–9.
68 He, Q., Zhang, H.J., Tian, Y., and Li, J.B. (2005) Comparative investigation of structure characteristics of mixed beta-lactoglobulin and different chain-length phophatidylcholine monolayer at the air/water interface. *Colloids Surf. A*, **257–258**, 127–131.
69 Miller, R., Fainerman, V.B., Leser, M.E., and Michel, M. (2004) Kinetics of adsorption of proteins and surfactants. *Curr. Opin. Colloid Interface Sci.*, **9**, 350–356.
70 Fainerman, V.B., Lucassen-Reynders, E.H., and Miller, R. (2003) Description of the adsorption behaviour of proteins at water/fluid interfaces in the framework of a two-dimensional solution model. *Adv. Colloid Interface Sci.*, **106**, 237–259.
71 Miller, R. (1991) Adsorption kinetics of macromolecules and macromolecule/surfactant mixtures at liquid interfaces. *Trends Polym. Sci.*, **2**, 47.
72 Ward, A.J.I. and Regan, L.H. (1980) Pendant drop studies of adsorbed films of bovine serum albumin. 1. Interfacial tensions at the isooctane–water interface. *J. Colloid Interface Sci.*, **78**, 389–394.
73 Tornberg, E. and Lundh, G. (1981) A study of the surface enlargement in the drop volume method and its relation to protein adsorption at a–w and o–w interfaces. *J. Colloid Interface Sci.*, **79**, 76–84.
74 Defeijter, J.A., Benjamins, J., and Veer, F.A. (1978) Ellipsometry as a tool to study adsorption behavior of synthetic and biopolymers at air–water interface. *Biopolymers*, **17**, 1760–1772.
75 Hansen, F.K. and Myrvold, R. (1995) The kinetics of albumin adsorption to the air/water interface measured by automatic axisymmetric drop shape analysis. *J. Colloid Interface Sci.*, **176**, 408–417.
76 Tripp, B.C., Magda, J.J., and Andrade, J.D. (1995) Adsorption of globular

proteins at the air/water interface as measured via dynamic surface tension – concentration dependence, mass transfer considerations, and adsorption kinetics. *J. Colloid Interface Sci.*, **173**, 16–27.
77 Murray, B.S. and Nelson, P.V. (1996) A novel Langmuir trough for equilibrium and dynamic measurements of air–water and oil–water monolayers. *Langmuir*, **12**, 5973–5976.
78 Murray, B.S. (1997) Equilibrium and dynamic surface pressure–area measurements on protein films at air–water and oil–water interfaces. *Colloids Surf. A*, **125**, 73–83.
79 Gonzalez, G. and MacRitchie, F. (1970) Equilibrium adsorption of proteins. *J. Colloid Interface Sci.*, **32**, 55.
80 Li, J.B., Zhang, Y., and L.L. Yan (2001) Multilayer formation on a curved drop surface. *Angew Chem Int Ed*, **40**, 891–894.
81 Dickinson, E. and Stainsby, G. (1982) *Colloids in Food*, Applied Science, London.
82 Fainerman, V.B., Miller, R., Ferri, J.K., Watzke, H., Leser, M.E., and Michel, M. (2006) Reversibility and irreversibility of adsorption of surfactants and proteins at liquid interfaces. *Adv. Colloid Interface Sci.*, **123–126**, 163–171.
83 Dickinson, E. and Matsumura, Y. (1994) Proteins at liquid interfaces: role of the molten globule state. *Colloids Surf. B*, **3**, 1–17.
84 Dickinson, E. and Matsumura, Y. (1991) Time-dependent polymerization of beta-lactoglobulin through disulfide bonds at the oil–water interface in emulsions. *Int. J. Biol. Macromol.*, **13**, 26–30.
85 Dill, K.A. (1990) Dominant forces in protein folding. *Biochemistry*, **29**, 7133–7155.
86 Norde, W. (1986) Adsorption of proteins from solution at the solid–liquid interface. *Adv. Colloid Interface Sci.*, **25**, 267–340.
87 Haynes, C.A. and Norde, W. (1994) Globular proteins at solid/liquid interfaces. *Colloids Surf. B*, **2**, 517–566.
88 Swaisgood, H.E. (1982) In *Developments in Dairy Chemistry*, vol. 1 (ed. P.F. Fox), Elsevier, London, p. 1.
89 Dickinson, E., Rolfe, S.E. and Dalgleish, D.G. (1990) Surface shear viscometry as a probe of protein–protein interactions in mixed milk protein films adsorbed at the oil–water interface. *Int. J. Biol. Macromol.*, **12**, 189–194.
90 Cornell, D.G. (1982) Lipid–protein interactions in monolayers – egg-yolk phosphatidic acid and beta-lactoglobulin. *J. Colloid Interface Sci.*, **88**, 536–545.
91 Cornell, D.G. and Patterson, D.L. (1989) Interaction of phospholipids in monolayers with beta-lactoglobulin adsorbed from solution. *J. Agric. Food Chem.*, **37**, 1455–1459.
92 Bos, M.A. and Nylander, T. (1996) Interaction between beta-lactoglobulin and phospholipids at the air/water interface. *Langmuir*, **12**, 2791–2797.
93 He, Q., Zhai, X.H., and Li, J.B. (2004) Direct visualization of the dynamic hydrolysis process of an L-DPPC monolayer catalyzed by phospholipase D at the air/water interface. *J. Phys. Chem. B*, **108**, 473–476.
94 Zhang, Y., An, Z.H., Cui, G.C., and Li, J.B. (2003) Stabilized complex film formed by co-adsorption of beta-lactoglobulin and phospholipids at liquid/liquid interface. *Colloids Surf. A*, **223**, 11–16.
95 Bos, M.A., Nylander, T., Arnebrant, T., and Clark, D.C. (1997) In *Food Emulsifiers and Their Applications* (eds G.L. Hasenhuettl and R.W. Hartel), Chapman & Hall, New York, p. 95.
96 Waite, M. (1987) *The Phospholipase*, Plenum Press, New York.
97 Rao, C.S. and Damodaran, S. (2002) Is interfacial activation of lipases in lipid monolayers related to thermodynamic activity of interfacial water? *Langmuir*, **18**, 6294–6306.
98 Dynarowicz-Latka, P., Dhanabalan, A., and Oliveira, O.N. (2001) Modern physicochemical research on Langmuir monolayers. *Adv. Colloid Interface Sci.*, **91**, 221–293.
99 Berg, O.G., Gelb, M.H., Tsai, M.D., and Jain, M.K. (2001) Interfacial enzymology: the secreted phospholipase A_2 paradigm. *Chem. Rev.*, **101**, 2613–2653.

100 Murakami, M. and Kudo, I. (2002) Arachidonate release and eicosanoid generation by group IIE phospholipase A_2. *J. Biochem.*, **131**, 285–292.

101 Balsinde, J., Winstead, M.V., and Dennis, E.A. (2002) Phospholipase A_2 regulation of arachidonic acid mobilization. *FEBS Lett.*, **531**, 2–6.

102 Nakanishi, M. and Rosenberg, D.W. (2006) Roles of cPLA$_2$ alpha and arachidonic acid in cancer. *Biochim. Biophys. Acta, Mol. Cell Biol. Lipids*, **1761**, 1335–1343.

103 Yedgar, S., Cohen, Y., and Shoseyov, D. (2006) Control of phospholipase A_2 activities for the treatment of inflammatory conditions. *Biochim. Biophys. Acta Mol. Cell Biol. Lipids*, **1761**, 1373–1382.

104 Li, J.B., Chen, Z.H., Wang, X.L., Brezesinski, G., and Möhwald, H. (2000) Dynamic observations of the hydrolysis of a DPPC monolayer at the air/water interface catalyzed by phospholipase A_2. *Angew. Chem. Int. Ed.*, **39**, 3059–3062.

105 Dahmen-Levison, U., Brezesinski, G., and Möhwald, H. (1998) Specific adsorption of PLA$_2$ at monolayers. *Thin Solid Films*, **327–329**, 616–620.

106 Wang, X.L., Zheng, S.P., He, Q., Brezesinski, G., Möhwald, H., and Li, J.B. (2005) Hydrolysis reaction analysis of L-alpha-distearoylphosphatidylcholine monolayer catalyzed by phospholipase A_2 with polarization-modulated infrared reflection absorption spectroscopy. *Langmuir*, **21**, 1051–1054.

107 Mendelsohn, R., Brauner, J.W., and Gericke, A. (1995) External infrared reflection-absorption spectrometry monolayer films at the air–water interface. *Annu. Rev. Phys. Chem.*, **46**, 305–334.

108 Dahmen-Levison, U., Brezesinski, G., and Möhwald, H. (1998) Structure studies of a phospholipid monolayer coupled to dextran sulfate. *Prog. Colloid Polym. Sci.*, **110**, 269–273.

109 Dahmen-Levison, U., Brezesinski, G., Möhwald, H., Jakob, J., and Nuhn, P. (2000) Investigations of lipid–protein interactions on monolayers of chain-substituted phosphatidylcholines. *Angew. Chem. Int. Ed.*, **39**, 2775–2778.

110 Wang, X.L., He, Q., Zheng, S.P., Brezesinski, G., Möhwald, H., and Li, J.B. (2004) Structural changes of phospholipid monolayers caused by coupling of human serum albumin: a GIXD study at the air/water interface. *J. Phys. Chem. B*, **108**, 14171–14177.

111 Zhai, X.L., Brezesinski, G., Möhwald, H., and Li, J.B. (2004) Thermodynamics and structures of amide phospholipid monolayers. *J. Phys. Chem. B*, **108**, 13475–13480.

112 Zhai, X.H., He, Q., Li, J.B., Brezesinski, G., and Möhwald, H. (2003) Self-organization of an L-ether-amide phospholipid in large two-dimensional chiral crystals. *ChemPhysChem*, **4**, 1355–1358.

113 Grainger, D.W., Reichert, A., Ringsdorf, H., and Salesse, C. (1990) Hydrolytic action of phospholipase-A_2 in monolayers in the phase-transition region – direct observation of enzyme domain formation using fluorescence microscopy. *Biochim. Biophys. Acta*, **1023**, 365–379.

114 Henon, S. and Meunier, J. (1991) Microscope at the Brewster angle direct observation of 1st-order phase transitions in monolayers. *Rev. Sci. Instr.*, **62**, 936–939.

115 Honig, D. and Mobius, D. (1991) Direct visualization of monolayers at the air–water interface by Brewster angle microscopy. *J. Phys. Chem.*, **95**, 4590–4592.

116 Zhai, X.H., Li, J.B., Brezesinski, G., He, Q., Möhwald, H., and Lai, L.H. (2003) Direct observations of the cleavage reaction of an L-DPPC monolayer catalyzed by phospholipase A_2 and inhibited by an indole inhibitor at the air/water interface. *ChemBioChem*, **4**, 299–305.

117 Zhai, X.H., Brezesinski, G., Möhwald, H., and Li, J.B. (2005) Impact of inhibiting activity of indole inhibitors on phospholipid hydrolysis by phospholipase A_2. *Colloids Surf. A*, **256**, 51–55.

118 Singer, W.D., Brown, H.A., and Sternweis, P.C. (1997) Regulation of

eukaryotic phosphatidylinositol-specific phospholipase C and phospholipase D. *Annu. Rev. Biochem.*, **66**, 475–509.
119 Basanez, G., Nieva, J.L., Goni, F.M., and Alonso, A. (1996) Origin of the lag period in the phospholipase C cleavage of phospholipids in membranes. Concomitant vesicle aggregation and enzyme activation. *Biochemistry*, **35**, 15183–15187.
120 Nielsen, L.K., Risbo, J., Callisen, T.H., and Bjørnholm, T. (1999) Lag-burst kinetics in phospholipase A_2 hydrolysis of DPPC bilayers visualized by atomic force microscopy. *Biochim. Biophys. Acta*, **1420**, 266–271.
121 He, Q. and Li, J.B. (2003) Dynamic and morphological investigation of phospholipid monolayer hydrolysis by phospholipase C. *Biochem. Biophys. Res. Commun.*, **300**, 541–545.
122 Lösche, M. and Möhwald, H. (1984) Fluorescence microscopy on monomolecular films at an air/water interface. *Colloid Surf.*, **10**, 217–224.
123 Gericke, A. and Huhnerfuss, H. (1994) IR reflection-absorption spectroscopy – a versatile tool for studying interfacial enzymatic processes. *Chem. Phys. Lipids*, **74**, 205–210.
124 Grandbois, M., Desbat, B., Blaudez, D., and Salesse, C. (1999) Polarization-modulated infrared reflection absorption spectroscopy measurement of phospholipid monolayer hydrolysis by phospholipase C. *Langmuir*, **15**, 6594–6597.
125 Cazzolli, R., Shemon, A.N., Fang, M.Q., and Hughes, W.E. (2006) Phospholipid signalling through phospholipase D and phosphatidic acid. *IUBMB Life*, **58**, 457–461.
126 Jenkins, G.M. and Frohman, M.A. (2005) Phospholipase D: a lipid centric review. *Cell Mol. Life Sci.*, **62**, 2305–2316.
127 Quarles, R.H. and Dawson, R.M.C. (1969) Hydrolysis of monolayers of phosphatidyl[Me-^{14}C]choline by phospholipase D. *Biochem. J.*, **113**, 697.
128 He, Q. and Li, J.B. (2003) "Lotus" domain formation by the hydrolysis reaction of phospholipase D to phospholipid monolayer. *Chin. Chem. Lett.*, **14**, 1199–1202.
129 Estrela-Lopis, I., Brezesinski, G., and Möhwald, H. (2001) Dipalmitoylphosphatidylcholine/phospholipase D interactions investigated with polarization-modulated infrared reflection absorption spectroscopy. *Biophys. J.*, **80**, 749–754.
130 Macdonal, B.F. and Barton, G. (1970) Lignans of western red cedar (*Thuja plicata* Donn.). 10. Gamma-thujaplicatene. *Can. J. Biochem.*, **48**, 3144.
131 Estrela-Lopis, I., Brezesinski, G., and Möhwald, H. (2000) Influence of model membrane structure on phospholipase D activity. *Phys. Chem. Chem. Phys.*, **2**, 4600–4604.
132 Zambonelli, C. and Roberts, M.F. (2005) Non-HKD phospholipase D enzymes: new players in phosphatidic acid signaling? *Prog. Nucleic Acid Res. Mol. Biol.*, **79**, 133–181.
133 Kirat, K.E., Prigent, A.F., Chauvet, J.P., Roux, B., and Besson, F. (2003) Transphosphatidylation activity of *Streptomyces chromofuscus* phospholipase D in biomimetic membranes. *Eur. J. Biochem.*, **270**, 4523–4530.
134 Geng, D., Chura, J., and Roberts, M.F. (1998) Activation of phospholipase D by phosphatidic acid – enhanced vesicle binding, phosphatidic acid Ca^{2+} interaction, or an allosteric effect? *J. Biol. Chem.*, **273**, 12195–12202.
135 Kirat, K.E., Besson, F., Prigent, A.F., Chauvet, J.P., and Roux, B. (2002) Role of calcium and membrane organization on phospholipase D localization and activity – competition between a soluble and an insoluble substrate. *J. Biol. Chem.*, **277**, 21231–21236.
136 Wagner, K. and Brezesinski, G. (2007) Phospholipase D activity is regulated by product segregation and the structure formation of phosphatidic acid within model membranes. *Biophys. J.*, **93**, 2373–2383.
137 Felnerova, D., Viret, J.F., Gluck, R., and Moser, C. (2004) Liposomes and virosomes as delivery systems for antigens, nucleic acids and drugs. *Curr. Opin. Biotechnol.*, **15**, 518–529.

138 Ropert, C. (1999) Liposomes as a gene delivery system. *Braz. J. Med. Biol. Res.*, **32**, 163–169.

139 Nasseau, M., Boublik, Y., Meier, W., Winterhalter, M., and Fournier, D. (2001) Substrate-permeable encapsulation of enzymes maintains effective activity, stabilizes against denaturation, and protects against proteolytic degradation. *Biotechnol. Bioeng.*, **75**, 615–618.

140 Ruysschaert, T., Germain, M., Gomes, J.F., Fournier, D., Sukhorukov, G.B., and Meier, W. (2004) Liposome-based nanocapsules. *IEEE Trans. Nanobiosci.*, **3**, 49–55.

141 Shao, Z., Mou, J., Czajkowsky, D.M., Yang, J., and Yuan, J.Y. (1996) Biological atomic force microscopy: what is achieved and what is needed. *Adv. Phys.*, **45**, 1–86.

142 Knoll, W. (1998) Interfaces and thin films as seen by bound electromagnetic waves. *Annu. Rev. Phys. Chem.*, **49**, 569–638.

143 Kung, L.A., Kam, L., Hovis, J.S., and Boxer, S.G. (2000) Patterning hybrid surfaces of proteins and supported lipid bilayers. *Langmuir*, **16**, 6773–6776.

144 Richter, R.P. and Brisson, A.R. (2005) Following the formation of supported lipid bilayers on mica: a study combining AFM, QCM-D, and ellipsometry. *Biophys. J.*, **88**, 3422–3433.

145 Giess, F., Friedrich, M.G., Heberle, J., Naumann, R.L., and Knoll, W. (2004) The protein-tethered lipid bilayer: a novel mimic of the biological membrane. *Biophys. J.*, **87**, 3213–3220.

146 Grandbois, M., Clausen-Schaumann, H., and Gaub, H.E. (1998) Atomic force microscope imaging of phospholipid bilayer degradation by phospholipase A_2. *Biophys. J.*, **74**, 2398–2404.

147 Wong, J.Y., Majewski, J., Seitz, M., Park, C.K., Israelachvili, J., and Smith, G.S. (1999) Polymer-cushioned bilayers. I. A structural study of various preparation methods using neutron reflectometry. *Biophys. J.*, **77**, 1445–1457.

148 Ross, M., Steinem, C., Galla, H.-J., and Janshoff, A. (2001) Visualization of chemical and physical properties of calcium-induced domains in DPPC/DPPS Langmuir–Blodgett layers. *Langmuir*, **17**, 2437–2445.

149 Sackmann, E. (1996) Supported membranes: scientific and practical applications. *Science*, **271**, 43–48.

150 Reimhult, E., Zäch, M., Höök, F., and Kasemoet, B. (2006) A multitechnique study of liposome adsorption on Au and lipid bilayer formation on SiO_2. *Langmuir*, **22**, 3313–3319.

151 Radtchenko, L., Sukhorukov, G.B., Leporatti, S., Khomutov, G.B., Donath, E., and Möhwald, H. (2000) Assembly of alternated multivalent ion/polyelectrolyte layers on colloidal particles. Stability of the multilayers and encapsulation of macromolecules into polyelectrolyte capsules. *J. Colloid Interface Sci.*, **230**, 272–280.

152 Moya, S., Donath, E., Sukhorukov, G.B., Auch, M., Bäumler, H., Lichtenfeld, H., and Möhwald, H. (2000) Lipid coating on polyelectrolyte surface modified colloidal particles and polyelectrolyte capsules. *Macromolecules*, **33**, 4538–4544.

153 Sukhorukov, B., Donath, E., Moya, S., Susha, A.S., Voigt, A., Hartmann, J., and Möhwald, H. (2000) Microencapsulation by means of step-wise adsorption of polyelectrolytes. *J. Microencapsul.*, **17**, 177–185.

154 Georgieva, R., Moya, S., Leporatti, S., Neu, B., Bäumler, H., Reichle, C., Donath, E., and Möhwald, H. (2000) Conductance and capacitance of polyelectrolyte and lipid–polyelectrolyte composite capsules as measured by electrorotation. *Langmuir*, **16**, 7075–7081.

155 Richter, R.P., Bérat, R., and Brisson, A.R. (2006) Formation of solid-supported lipid bilayers: an integrated view. *Langmuir*, **22**, 3497–3505.

156 Ariga, K. (2004) Silica-supported biomimetic membranes. *Inc. Chem. Rec.*, **3**, 297–307.

157 Plant, A.L. (1999) Supported hybrid bilayer membranes as rugged cell membrane mimics. *Langmuir*, **15**, 5128–5135.

158 Sackmann, E. and Tanaka, M. (2000) Supported membranes on soft polymer

cushions: fabrication, characterization and applications. *Trends Biotechnol.*, **18**, 58–64.

159 Lösche, M., Schmitt, J., Decher, G., Bouwman, W.G., and Kjaer, K. (1998) Detailed structure of molecularly thin polyelectrolyte multilayer films on solid substrates as revealed by neutron reflectometry. *Macromolecules*, **31**, 8893–8906.

160 Cassier, T., Sinner, A., Offenhäuser, A., and Möhwald, H. (1999) Homogeneity, electrical resistivity and lateral diffusion of lipid bilayers coupled to polyelectrolyte multilayers. *Colloids Surf. B*, **15**, 215.

161 Kügler, R. and Knoll, W. (2002) Polyelectrolyte-supported lipid membranes. *Bioelectrochemistry*, **56**, 175–178.

162 Spinke, J., Liley, M., Guder, H.J., Angermaier, L., and Knoll, W. (1993) Molecular recognition at self-assembled monolayers – the construction of multicomponent multilayers. *Langmuir*, **9**, 1821–1825.

163 Stamm, M., Hüttenbach, S., and Reiter, G. (1991) TOREMA – a neutron reflectometer at Julich. *Physica B*, **173**, 11–16.

164 Jenkins, A.T.A., Bushby, R.J., Boden, N., Evans, S.D., Knowles, P.F., Liu, Q.Y., Miles, R.E., and Ogier, S.D. (1998) Ion-selective lipid bilayers tethered to microcontact printed self-assembled monolayers containing cholesterol derivatives. *Langmuir*, **14**, 4675–4678.

165 Soumpasis, D.M. (1983) Theoretical analysis of fluorescence photobleaching recovery experiments. *Biophys. J.*, **41**, 95–97.

166 Delajon, C., Gutberlet, T., Steitz, R., Möhwald, H., and Krastev, R. (2005) Formation of polyelectrolyte multilayer architectures with embedded DMPC studied *in situ* by neutron reflectometry. *Langmuir*, **21**, 8509–8514.

167 Wagner, M.L. and Tamm, L.K. (2001) Reconstituted syntaxin1A/SNAP25 interacts with negatively charged lipids as measured by lateral diffusion in planar supported bilayers. *Biophys. J.*, **61**, 266–275.

168 Knoll, W., Frank, C.W., Heibel, C., Naumann, R., Offenhäusser, A., Rühea, J., Schmidt, E.K., Shen, W.W., and Sinner, A. (2000) Functional tethered lipid bilayers. *Rev. Mol. Biotechnol.*, **74**, 137–158.

169 Troutier, A.L. and Ladavière, C. (2007) An overview of lipid membrane supported by colloidal particles. *Adv. Colloid Interface Sci.*, **133**, 1–21.

170 Johnston, A.P.R., Zelikin, A.N., Lee, L., and Caruso, F. (2006) Approaches to quantifying and visualizing polyelectrolyte multilayer film formation on particles. *Anal. Chem.*, **78**, 5913.

171 Ge, L.Q., Möhwald, H., and Li, J.B. (2003) Biointerfacing polyelectrolyte microcapsules. *Chem. Phys. Chem.*, **4**, 1351–1355.

172 Decher, G. and Hong, J.D. (1991) Buildup of ultrathin multilayer films by a self-assembly process. 1. Consecutive adsorption of anionic and cationic bipolar amphiphiles on charged surfaces. *Makromol. Chem. Macromol. Symp.*, **46**, 321–327.

173 Caruso, F., Donath, E., and Möhwald, H. (1998) Influence of polyelectrolyte multilayer coatings on Forster resonance energy transfer between 6-carboxyfluorescein and rhodamine B-labeled particles in aqueous solution. *J. Phys. Chem. B*, **102**, 2011–2016.

174 Ge, L.Q., Möhwald, H., and Li, J.B. (2003) Phospholipase A_2 hydrolysis of mixed phospholipid vesicles formed on polyelectrolyte hollow capsules. *Chem. Eur. J.*, **9**, 2589–2594.

175 He, Q., Tian, Y., Möhwald, H., and Li, J.B. (2009) Biointerfacing luminescent nanotubes. *Soft Matter*, **5**, 300–303.

2
Layer-by-Layer Assembly of Biomimetic Microcapsules

2.1
Introduction

Molecular biomimetics (i.e., biomimetics on the molecular level) can provide novel design approaches and manufacturing processes [1–5]. For instance, biomolecules such as proteins, lipids, DNA, RNA, and so on, as well as the structures and forms that these molecules assemble, biomembranes, the nucleus, mitochondria, endosomes, and others, serve as rich sources of ideas for scientists or engineers interested in developing novel biomimetic materials for innovations in biophysical research and the biomedical field. In fact, molecular biomimetics has proved very useful in the design and fabrication of new functional structured materials on the micro- and nanoscale, and they display superior physical and biological properties in comparison with synthetic composites [6–8].

The cell is the structural and functional unit of all known living organisms in natural biological systems. All cells have an enveloping membrane. The cell membrane or plasma membrane with a bilayer structure formed by the self-assembly of lipid molecules is a flexible barrier that serves to separate the cell's interior from its environment, regulate what moves in and out (selectively permeable), and maintain the electric potential of the cell. This biological membrane is composed of a closed lipid bilayer structure (i.e., compartmentalization) embedded with various membrane proteins. Compartmentalization is one of the key architectural principles of living cells that distinguishes them from less-evolved forms of life [9]. The compartments exchange nutrients or metabolites through a variety of processes including vesicle budding, membrane fusion, membrane protein-assisted permeation, and movement. In principle, the proper functioning of cells requires precise control over metabolic reactions, which is achieved by segregation of biomolecules in compartments and exchange through boundaries by means of selective transport processes. To mimic the structure and function of cells, several model biomembrane systems, such as lipid monolayers, liposomes, and supported lipid bilayers, have been employed in order to understand the properties of biological membranes. Conventionally, from the viewpoint of biophysical research, liposomes should thus be a good cell mimic because compartmentalization is a fundamental requirement for the reproduction of the natural environment. However, the limitation of size and stability of the assembled liposome complexes results in difficulty understanding and analyzing them [9, 10]. On the other hand, one of the most promising applications of these biomimetic hollow vesicles is to develop delivery systems. However, these hollow systems often have limited applications owing to their own instability or a specific chemistry required for their fabrication. In particular, many bioactive compounds cannot, from a practical aspect, effectively be delivered to a specific site in a living organism because of the lack of specificity, metabolic stability, or bioavailability [11]. Therefore, much attention has recently been given to build up robust biomimetic membrane systems with uniform size, and good mechanical and chemical stability where conformational freedom for proteins is as large as possible [12, 13]. These new systems should also contain a suitable compartment so that the natural environment of membrane-bound proteins can be recreated. Up to now, one of the better candidates for a cell mimic has been proved to be polymer-cushioned lipid membrane systems. In fact, biomembranes in living organisms are supported by a polymer network – the cytoskeleton (or in the case of bacteria and plants, the cell wall). For this reason, polymer-cushioned lipid membranes should be a promising biomimetic membrane system.

2.2
Layer-by-layer Assembly of Polyelectrolyte Multilayer Microcapsules

2.2.1
General Aspects

In the beginning of the 1990s the layer-by-layer (LbL) technique introduced by Decher and Hong was originally used for sequential adsorption of oppositely charged polymers, polyelectrolytes, on a charged surface [14]. When a polyelectrolyte is adsorbed on a charged surface, the charge of the polymer overcompensates, leading to a reversal of the surface charge, thereby promoting the adsorption of the next oppositely charged polyelectrolyte. A stable film with thickness and roughness controllable in the nanometer range can be obtained [15]. Depending predominantly on electrostatic interactions, it can be applied with almost any multiply charged entities like synthetic and natural polymers and oligomers, multivalent ions and dyes, proteins, and inorganic and organic particles [16, 17]. Apart from electrostatic interactions, other interactions such as hydrogen bonding, covalent bonding, biospecific interactions, and stereocomplex formation have also been used to build up these multilayer film systems. Thus, multifunctional organic and composite films can be made by simple adsorption procedures [18, 19]. In 1998, Moehwald *et al.* extended the LbL technique to polymeric capsules by depositing polyelectrolytes onto charged colloidal particles as templates (Figure 2.1) [20]. For the fabrication of hollow polyelectrolyte capsules, a template is used that can later be removed. Typical adsorption conditions used to form polyelectrolyte layers on

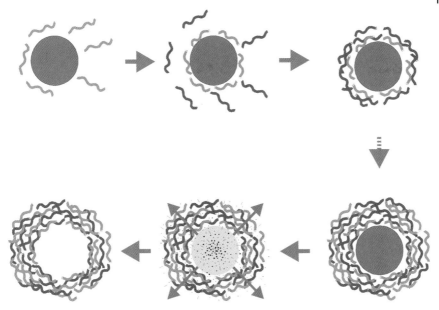

Figure 2.1 Schematic illustration of the polyelectrolyte deposition process for PSS/PAH capsules and of the subsequent MF core decomposition. The initial steps involve stepwise film formation by repeated exposure of the colloid MF template to alternatively charged polyelectrolytes. The excess polyelectrolyte is removed by cycles of centrifugation and washing before the next layer is deposited. After the desired number of layers has been deposited, the coated particles are exposed to HCl pH 1. The MF core immediately decomposes. Finally, after washing, a suspension of polyelectrolyte hollow shells is obtained (Reprinted with permission from [20]. © 1998, Wiley).

colloidal particles are taken from the planar film assembly. Since an excess of polyelectrolyte is needed, the total particle surface in the batch must be evaluated from the colloid concentration and diameter. The drying process often applied in planar film assembly between the deposition of two subsequent layers cannot be performed with particles. LbL assembly of polyelectrolyte onto colloidal particles can be performed in two different ways: either the concentration of polyelectrolyte added at each step is just sufficient to form a saturated layer or the adsorption is carried out in an excess of polyelectrolyte. In the latter case, the excess polyelectrolyte molecules must be removed before adding the next polyelectrolyte to avoid formation of complexes in solution. This separation can be achieved either by centrifugation or filtration [21, 22].

Several methods have been applied to check the layer growth of polyelectrolytes onto colloidal suspensions. For polyelectrolytes, electrophoresis yields a complete charge reversal at each step as measured by the ζ (zeta) potential [23]. For ampholytes or weakly charged molecules, the charge can alternate between a weak and strong positive (or negative) ζ potential. With single particle light scattering, the adsorption of each polyelectrolyte layer can be detected separately [21]. In addition, this method can discriminate between single particles and aggregates. Recently, small-angle neutron scattering measurements of poly(allylamine hydrochloride) (PAH)/poly(sodium 4-styrenesulfonate) (PSS) multilayers adsorbed on deuterated latex particles allowed the direct determination of the thickness of the polyelectrolyte multilayer on coated colloids in water [24]. Transmission electron microscopy (TEM) was applied to follow the polyelectrolyte layer growth, especially for very small particles [25]. After removal of the core, atomic force microscopy (AFM) enabled the determination of the surface profile, the layer thickness of dried and collapsed capsules, as well as their diameter, shape, and filling [26].

Typically monodispersed weak cross-linking melamine formaldehyde (MF) particles are synthesized for use as templates. The PAH/PSS multilayer deposition can be performed by repeated exposure of the colloid particles to oppositely charged polyelectrolyte solutions. The excess polyelectrolyte is removed by cycles of centrifugation and washing. After the formation of multilayers, the cores are exposed in a 0.1 M HCl solution. Cores will be dissolved and the PAH/PSS hollow capsules are formed. However, the MF capsules proved difficult to remove after the formation of the multilayer and the in case of polystyrene microparticle templates an osmotic pressure could be built up in the capsule due to the fast dissolution of the polymer core, which could destroy the polymeric shell. To overcome these problems inorganic substrates such as $CaCO_3$, $MnCO_3$, and $CdCO_3$ are now used more frequently [27]. These inorganic carbonates have the advantage that upon dissolution with an ethylenediamine tetraacetate (EDTA) solution the metal ions are complexed and can pass through the membrane of the polyelectrolyte shell. These capsules are known for their permeability of molecules with a molecular weight below 5 kDa and should therefore have no osmotic stress.

Precipitates of enzymes were used as templates targeting catalysis. In this case, the templates are not spherical like latex particles. Thus, polyelectrolyte multilayers coated on enzyme crystals follow the morphology of the template while switching from rectangular to spherical after solubilization of the enzyme crystal due to the establishment of osmotic pressure in parallel with the template dissolution [28]. In addition, one must carefully control the conditions. Catalase microcrystals were coated with multilayers at pH 5.0 and 4 °C to ensure the catalase crystal surface was positively charged and not soluble. In addition, the aggregates of chymotrypsin ranged from 100 to 300 nm [29]. The precipitation of protein can be prepared by mixing the enzyme and saline solutions. The multilayer shells considerably protect the enzymes and about 70% of the activity of the encapsulated chymotrypsin remained. However, if a high-molecular-weight polyelectrolyte is used to construct the multilayer coating, the low permeability confines the encapsulated enzyme diffusion from inside to outside of the capsules, thus limiting the application.

Wang *et al.* exploited mesoporous silica (MS) as a template to prepare nanoporous polymer spheres. Poly(acrylic acid) (PAA) together with PAH was used as the polyelectrolyte pair [30]. Furthermore, they also studied the infiltration of PAA into amine-functionalized MS particles as a function of molecular weight, nanopore size, and solution conditions [31]. Tong *et al.* used the cross-linking method to selectively form amide bonds between the PAA and the primary amine groups of the template, thereby stabilizing the adsorbed PAA molecules [32, 33]. The particles were then exposed to PAH and cross-linking was again performed to bond the carboxylic acid moieties of PAA and the amine groups of PAH. The LbL techniques described above use solid particles in an aqueous medium. A novel development is LbL deposition on a liquid core. Grigoriev *et al.* have recently described a general method for the encapsulation of a broad range of emulsions comprised of various hydrophobic substances and oil-soluble compounds [34]. To an emulsion of didodecyldimethylammonium bromide (DODAB), chloroform, and dodecane in water was gently added a polyelectrolyte solution. After deposition of a layer, a washing step was carried out, before the next layer was deposited, resulting in microcapsules. Encapsulation of an oil was also performed using an ultrasonic probe in combination with the LbL technique [35]. A hydrophobic drug was dissolved in soybean oil and transferred to a polyglutamate solution containing an emulsifier. The emulsion was sonicated to form spheres of the desired size and layers were deposited in the same manner as described before.

2.2.2
Permeation and Mechanical Properties of LbL Microcapsules

Permeation could be measured as given in Figure 2.2 [28]. In this case the left fluorescence micrograph shows that the fluorescently labeled protein does not

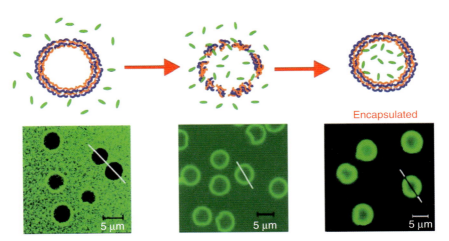

Figure 2.2 Permeability of MF-derived capsules for FITC–dextran. The bottom row shows confocal fluorescent micrographs of MF derived capsules in FITC–dextran (77 kDa) solution at pH 8 (left), pH 6 (middle), and returning to pH 8 and removing the external FITC–dextran by centrifugation (right). The top row shows the interpretation of the images with reversible pore formation at pH 6 and encapsulation of macromolecules at pH 8 (Reprinted with permission from [28]. © 2001, American Chemical Society).

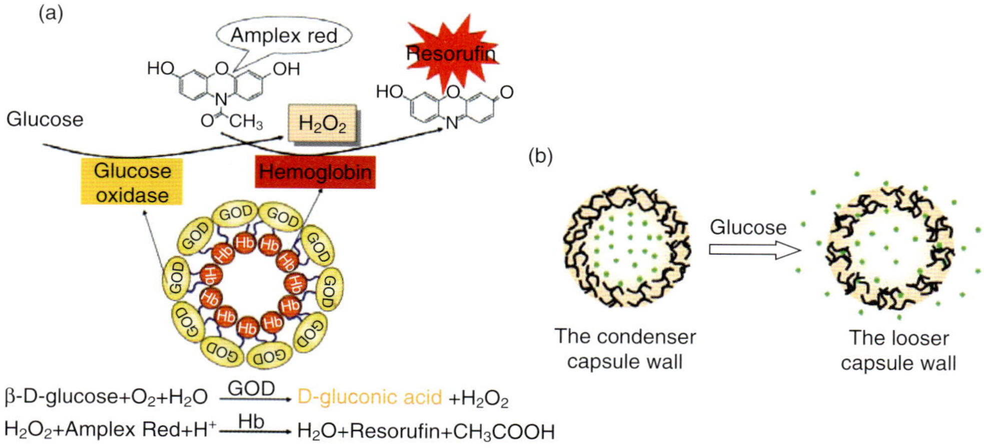

Figure 2.3 (a) Schematic illustration of coupled enzymatic process based on GOD and Hb coimmobilized as LbL microcapsules components. (b) Illustration of glucose-stimulated FITC–dextran release (green circles). The consumption of glucose enhanced the release of the encapsulated materials (Reprinted with permission from [39]. © 2009, American Chemical Society).

β-D-glucose + O_2 + H_2O $\xrightarrow{GOD}$ D-gluconic acid + H_2O_2

H_2O_2 + Amplex Red + H^+ $\xrightarrow{Hb}$ H_2O + Resorufin + CH_3COOH

penetrate the wall. Decreasing the pH from 8 to 6, the protein can penetrate as indicated by the fluorescence from inside the capsules (middle image). Going back to pH 8 and removing the protein outside, one keeps the once penetrated macromolecules inside (right image); hence, we have pH switchable permeation, which in addition is reversible. This example shows that the permeability can be switched by varying the intermolecular interactions. Other parameters to achieve this have been treatment by solvents (acetone, ethanol) and changing the salt concentration [36]. Incorporating thermosensitive copolymers led to temperature-dependent permeation, but this was irreversible [37]. Also, dye aggregates could be incorporated, enabling photoinduced permeation for dextrans of a specific size, but again this process was not reversible [38]. Our group fabricated glucose-sensitive protein multilayer capsules by alternate assembly of hemoglobin (Hb) and glucose oxidase (GOD) with glutaraldehyde (GA) as cross-linker (Figure 2.3) [39, 40]. GOD catalyzes the oxidation and hydrolysis of β-D-glucose into gluconic acid and H_2O_2. Hb can catalyze the reduction of H_2O_2 with its certain intrinsic peroxidase activity and nonfluorescent Amplex Red was oxidized by H_2O_2 into resorufin (a fluorescence dye). Thus, the system involves two enzymatic catalysis reactions and it offers two advantages. One is that the fluorescence of resorufin makes it convenient to monitor the reactions process, which may facilitate the development of an approach to design fluorescence sensors. The other is that the consumption of glucose enhanced the permeability of the capsule wall and increased the release of the encapsulated drugs (Figure 2.3b), which is highly attractive for the fabrication of glucose-responsive release systems.

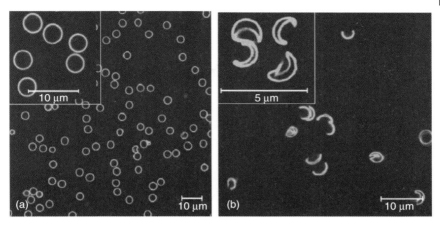

Figure 2.4 Confocal images of fluorescently labeled polyelectrolyte microcapsules before (a) and after (b) the addition of polyelectrolyte in the outer medium. While the capsules in (a) are perfectly spherical, buckling has occurred in (b) and the capsules are folded into a cup shape (Reprinted with permission from [41]. © 2001, American Chemical Society).

Additionally, the mechanical properties of polyelectrolyte capsules in aqueous solution or a dried state play an important role in many applications. In general, a higher mechanical stability can be achieved for LbL films than for fluid membranes or liposomes. Moreover, the mechanical properties can be varied over a wide range by using different compositions of polyelectrolytes and by increasing the number of deposited layers. Cross-linking of the noncovalently bound polyelectrolyte multilayers can improve capsule stability and cause a transition from a viscous to an elastic response. Gao *et al.* proposed a simple method to determine moduli of microcapsule walls [41]. Figure 2.4 shows the effects of fluorescently labeled capsules on increasing the polyelectrolyte (PSS) concentration from left to right. Since PSS does not penetrate the wall, it creates an osmotic pressure that can also be measured by osmometry and above a certain threshold this creates shape instability. This threshold can be determined by counting the fraction of deformed capsules as a function of polyelectrolyte concentration.

2.3
Biointerfacing Polyelectrolyte Microcapsules – A Multifunctional Cargo System

Our group and other researchers recently develop a new type of biomimetic microcapsule – lipid-coated polyelectrolyte microcapsules or polyelectrolyte multilayer-cushioned liposomes. This new biomimetic system is based on the above-mentioned liposome and LbL-assembled polyelectrolyte multilayer (Figure 2.5) [42–52]. Polyelectrolyte multilayer-supported liposome systems or lipid bilayer-coated

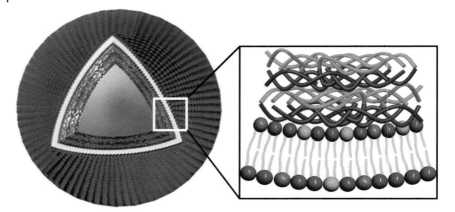

Figure 2.5 Representation of a lipid bilayer-coated LbL microcapsule.

LbL-assembled capsules also have these properties accordingly. Thus, one cannot only conveniently tune their morphological properties, including the exterior shape, interior space, and shell structure, but also realize their multifuctionalization. Polyelectrolyte multilayer-supported liposome systems should be useful for the understanding of the principles of the interaction of membranes with biopolymers such as proteins, and enable the design and application of new biomimetic structured materials. These lipid-modified polymer microcapsules could be considered as an ideally supported biomimetic membrane system to mimic the real cell membrane.

2.3.1
Lipid Bilayer-Modified Polyelectrolyte Microcapsules

Polyelectrolyte multilayer-supported liposomes or lipid bilayer-coated polyelectrolyte multilayer capsules could be fabricated through the conversion of liposomes into lipid bilayers to cover the capsule surface in analogy to the cell membrane. These lipid-modified polymer microcapsules should be an ideal supported biomimetic membrane system to mimic a real cell membrane. This new hybrid system also enables the design and application of new biomimetic structured materials.

Commonly, the first step of this biomimetic membrane fabrication process consists of preparing a hollow polyelectrolyte multilayer capsule by the alternating LbL procedure templated on spherical particles [42]. In most cases, PSS and PAH were used as wall components. The protocol used for the lipid coating was mainly based on electrostatic attraction between the last polyelectrolyte layer and unilamellar vesicles composed of lipids with opposite charges. Vesicles were incubated with the capsule dispersion in order to let adsorption occur and the samples were then centrifuged to remove nonadsorbed lipids. These assemblies were characterized by means of various spectroscopic and microscopic techniques to evidence the successful coating of lipid bilayers on the surface of capsules. Moya et al. followed the ζ potential change of the surface of polyelectrolyte capsules to identify the effective lipid adsorption [53]. The inversion of the charge sign of the capsule surface after contact with oppositely charged vesicles proved the surface modification by lipids. The sign inversion was less pronounced for formulations containing zwitterionic lipids owing to their neutral polar head. To further demonstrate lipid adsorption on capsules, a fluorescent lipid (N-(7-nitro-2,1,3-benzoxadiazol-4-yl) (NBD)-phosphocholine) was employed and assemblies were observed by confocal laser scanning microscopy (CLSM). The images showed the presence of a fluorescent ring around capsules and, as the fluorescence intensity distribution was homogeneous, lipid coverage was considered as uniform. To confirm the coating quality, these fluorescent species were analyzed by flow cytometry. The results show that whatever lipid formulation, an intense and thin fluorescence peak was obtained. The authors concluded on a homogeneous adsorption and explained that in the case of incomplete lipid coverage, peaks would have been larger due to additional signals in the low fluorescence intensity values.

The existence of a phase transition temperature – a typical characteristic of a lipid membrane – can also be used to estimate whether lipid layer on the capsules forms an organized layer structure or not. Differential scanning calorimetry (DSC) measurements can provide information on the phase of the adsorbed lipids by measuring the phase transition enthalpy [54]. Some additional information on the extent of coupling of the lipids to the polyelectrolyte support could thus be given. After lipid absorption another layer of PAH was deposited to provide a symmetric environment to the lipid layers. The existence of a transition peak confirmed that lipids formed an ordered phase and did not occur as single molecules adsorbed on supports. For the anionic lipids adsorbed on capsules, the phase transition temperature was 8 °C lower than that for the same lipids in vesicles. This variation was attributed to a strong coupling interaction between lipids and polyelectrolyte able to disturb lipid organization via electrostatic interactions.

The surface roughness of lipid layer-coated polyelectrolyte capsules was evaluated by AFM. The images show that the surface is almost perfectly smooth and the root mean square of the height variations ("rms value") obtained is less than 1 nm. The surface of polyelectrolyte layers prior to the lipid deposition presented a considerable roughness and the rms value was about 7 nm accordingly. After incubation of vesicles and capsules, it is clearly obvious that the lipid coverage has considerably smoothed out the roughness of the polymeric support. The height variations are reduced by a factor of 2. Obviously, the lipid coating reduced surface irregularities and uniformed the relief of the assemblies.

Additionally, Moya et al. carried out freeze-fracture electron microscopy experiments to prove if the lipids formed a bilayer structure when adsorbed onto the capsules [54]. Freeze-fracture micrographs of bare particles and polyelectrolyte-coated particles indicated that fracture took place through the hydrophobic polystyrene interior. On the contrary, assemblies surrounded by lipids fractured through the weakest part, which is the mid-plane of the adsorbed lipid bilayer. The micrographs confirm mainly continuous lipid coverage and the presence of a lipid bilayer structure. Afterward, the Förster energy transfer between two fluorescent probes was quantitatively assessed to derive the thickness of lipid layers on the capsules. Briefly, lipid layers were sandwiched between two polycation dye-labeled PAH layers – one being marked with fluorescein (donor molecule) and the other one with rhodamine (acceptor molecule). Considering that the thickness of one polyelectrolyte layer is 1.5–2 nm, the distance between both probes was estimated to be around 4–5 nm in the case of lipid coverage.

To test the successful coating of lipid layers, permeability assays were carried out on the lipid/polyelectrolyte capsules by our group [44]. The fluorescent probe 6-carboxyfluorescein (6-CF, insoluble in lipid membrane) was incubated with the capsules and observations by CLSM allowed us to evidence that the lipid envelope decreases the capsule permeability to molecules by a factor of 10^3. Furthermore, these systems remained impermeable during at least 6 months (at 4 °C). Another strategy consisted in extemporaneously incorporating 6-CF to reference capsule solutions (i.e., without lipids) and partially or completely lipid-covered capsules. The study by confocal microscopy showed that the probe entered into reference capsules and partially-covering the capsules, whereas a continuous lipid shell avoided this phenomenon and led to perfectly impermeable hollow systems. Subsequently, an enzyme, phospholipase A_2 (PLA$_2$), was added into a lipid-coated capsule suspension to tune the permeability (Figure 2.6) [42]. PLA$_2$ exists extensively in living organisms. It can stereoselectively hydrolyze the *sn*-2 ester linkage of enantiomeric L-phospholipids, such as L-α-dipalmitoylphosphatidylcholine (L-DPPC), to release fatty acids and lysophospholipids [55]. The hydrolytic product, lysophospholipid, is water-soluble and may leave from the interface to the solution,

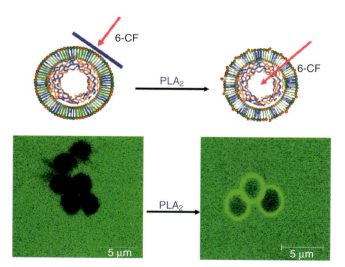

Figure 2.6 Schematic description of PLA$_2$-catalyzed lipid-modified polyelectrolyte multilayer microcapsule and corresponding CLSM images (Reprinted with permission from [42]. © 2003, Wiley).

which can lead to the loss of lipid layer material. In other words, the enzyme creates defects in lipid layers and the permeability of lipid layer-coated capsules can be adjusted accordingly. As a consequence, by formulating adequate lipid mixtures, this system can offer promising opportunities for controlled delivery applications.

2.3.2
Formation of Asymmetric Lipid Bilayers on the Surface of LbL-Assembled Capsules

In biological systems, a key feature of cell plasma membranes is the asymmetric distribution of phospholipids, which controls important cellular processes such as signal transduction and fusion. It has been found that the aminophospholipids, such as phosphatidylserine, are enriched on the inner leaflet, while the outer leaflet has more phosphatidylcholine. Although asymmetric vesicles with inner and outer monolayers composed of different lipids have been prepared, size distributions of the asymmetric vesicles are relatively broad.

Recently, Katagiri and Caruso reported the preparation of monodisperse vesicles comprising asymmetric lipid bilayers supported on colloidal particles coated with LbL-assembled polyelectrolyte multilayers [56]. In their report, unilamellar lipid bilayers on polyelectrolyte multilayer microcapsules were formed (Figure 2.7). Briefly, polyelectrolyte microcapsules with a narrow size distribution were first fabricated via LbL assembly. Next, an "inner" lipid monolayer from organic solution was deposited onto the oppositely charged polyelectrolyte microcapsules. An "outer" lipid monolayer coating was formed by incubating small unilamellar vesicles with the lipid monolayer-coated colloids in water. Poly(diallyldimethylammonium chloride) (PDDA) and PSS were employed as the polycation and polyanion, respectively, while dimethyldioctadecylammonium bromide (DDAB) and dihexadecyl phosphate (DHP), sodium salt, were used as the cationic and anionic lipids, respectively. Finally, asymmetric lipid bilayer-modified polyelectrolyte capsules or vesicular particles were obtained. Similarly, multilamellar lipid bilayers on polyelectrolyte multilayer microcapsules can be formed by repetition of these three steps.

Multilayer film formation followed by microelectrophoresis shows the alternating ζ potential as a function of PSS/PDDA layers and DDAB/DHP asymmetric layers. The ζ potential measurements confirmed the assembled system with DHP as the inner leaflet and DDAB as the outer leaflet. The quartz crystal microbalance (QCM) was employed to follow the assembly process. According to the Sauerbrey equation, the frequency decreases proportionally with an increase in mass deposited on the crystal surface. The authors hoped that the amount of polyelectrolyte and lipid deposited with each adsorption step could be quantified via the QCM technique. The QCM measurement was performed by recording values of frequency change at each step of the assembly of the PSS/PDDA/DDAB$_{inner}$/DHP$_{outer}$ system. The observation shows a regular film growth for this system and is in good agreement with that calculated for a hexagonally packed lipid monolayer of DDAB or DHP. Both DDAB and DHP were mainly deposited as monolayers according to their analysis.

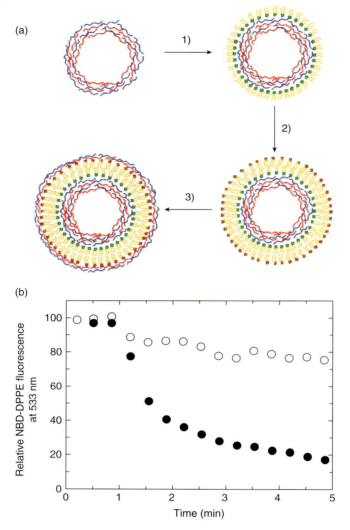

Figure 2.7 (a) Illustration of the assembly of an asymmetric lipid bilayer membrane (i.e., a lipid membrane where the inner and outer layers of the bilayer comprise different lipid molecules) on an LbL-assembled polyelectrolyte microcapsule, followed by several additional polyelectrolyte layers. (b) Relative NBD-DPPE fluorescence at 533 nm as a function of time for MF particles coated with a polyelectrolyte-supported asymmetric lipid bilayer where DHP (containing 3 mol% NBD-DPPE) forms the inner (open circles) or outer (closed circles) layer of the bilayer membrane. In both cases, NBD-DPPE is embedded in the DHP layer and DDAB forms the alternate lipid layer. (Reprinted with permission from [56]. © 2005, Wiley).

Fluorescence microscopy was applied to examine the deposition of DDAB and DHP layers onto PSS/PDDA-coated particles. NBD-dipalmitoylphosphatidylethanolamine (DPPE) was incorporated in the anionic DHP layer as the anionic lipid dye probe. Homogeneous fluorescence originating from the surface of the polyelectrolyte/lipid-coated particles proves the presence and homogeneous distribution of lipids on the particle surface, even after removal of the templating core. Similarly, fluorescence spectroscopy was used to follow the construct of PSS/PDDA/PSS/(DDAB/DHP/PDDA/PSS)$_n$ multilayers ($n = 1, 2,$ or 3) in water. The linear increase in the maximum emission intensity of NBD-DPPE observed for the $n = 1, 2,$ and 3 coatings indicates the deposition of multiple asymmetric bilayers.

In order to prove that the lipid membrane is indeed asymmetric (i.e., that the inner and outer layers of the bilayer comprise different lipid molecules), fluorescence quenching assays were carried out to determine the distribution of probe lipids between the inner and outer leaflets of the bilayer membrane. NBD-DPPE was again used as a probe lipid in the DHP monolayer. Sodium hydrosulfite ($Na_2S_2O_4$) was adopted as a fluorescence quencher for the NBD-derivative probe. The fluorescence emission intensity measurement before and after addition of $Na_2S_2O_4$ solution shows that the addition of the quencher to the dispersion indeed causes a reduction in the emission of only the probe on the outer monolayer of the membrane. They also found that the relative fluorescence intensity is a function of fluorescence quenching time and thus confirmed the formation of an asymmetric bilayer.

The fluorescence quenching assay also demonstrates that the asymmetry of the lipid bilayers can be stabilized for at least 72 h, revealing that the lipid flip-flop phenomenon does not occur in this asymmetric lipid membrane system. Additionally, no change of the emission intensity of NBD-DPPE after the addition of Triton X-100 shows the high morphological stability of the polyelectrolyte-supported lipid membrane. The authors ascribe this to complexation between the lipid molecules in each close-packed monolayer and the polyelectrolyte molecules in the adjacent multilayers. The polyelectrolyte multilayers on either side of the asymmetric lipid bilayer membrane play an important role in preserving the membrane's stability. Monodisperse polyelectrolyte-supported vesicles with asymmetric lipid bilayers are closely related to those of the original template core particles, thereby providing a means to control the size.

2.3.3
Assembly of Lipid Bilayers on Covalently LbL-Assembled Protein Capsules

An extension of electrostatic LbL-assembled multilayer films exploits other nonelectrostatic interactions to facilitate film assembly. Among them, the use of covalent bonds to assemble LbL multilayer films can provide significant advantages compared to traditional electrostatic assembly [57]. In particular, they have high stability due to the covalent bonds formed and therefore do not disassemble with changes in pH or ionic strength. For instance, Hb is a conjugated iron-protein compound in red blood cells, and transports oxygen, carbon dioxide, and nitric

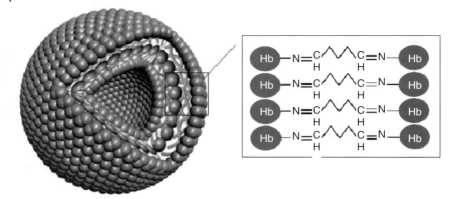

Figure 2.8 Schematic representation of the assembled Hb microcapsules via covalent LbL deposition (Reprinted with permission from [58]. © 2007, Elsevier).

oxide. Hb plays an important role in vital activities. Therefore, the fabrication of Hb capsules could imitate in some sense its structural function in the living system, and may help us to understand its properties and further make protein-based biodevices. Our group has recently prepared Hb capsules by using a covalent LbL technique [58]. In this work, GA is used as a chemical cross-linker because it has less effect on the protein activity. Poly(ethylenimine) (PEI) was first adsorbed on template particles to produce an amino-functionalized surface. Then, the GA and Hb were alternately adsorbed. Figure 2.8 shows a schematic representation of a covalently cross-linked Hb capsules. Direct information on the capsule formation can be obtained from the measurements by TEM and CLSM. The results show that the wall thickness of capsules can be controlled by the adsorption cycles of alternate GA/Hb. The UV-Vis spectra of GA/Hb capsules show the absorption band of heme at 411 nm, indicating that Hb essentially remains in the capsules. Cyclic voltammetry and potential-controlled amperometric measurements confirm that cross-linked Hb capsules keep their heme electroactivity and are not denatured. The typical amperometric response toward the successive addition of H_2O_2 shows the electrocatalytic property of GA/Hb capsules. The permeability of the assembled GA/Hb microcapsules was tested by using fluorescein isothiocyanate (FITC)–dextran of different molecular weights as fluorescent probes. The results show that the $(GA/Hb)_5$ capsules are impermeable to FITC–dextran with molecular weights of 2000 and 500 kDa, while FITC–dextran with molecular weights below 70 kDa can partly, and even completely, permeate into the capsule interior. In comparison with traditional $(PAH/PSS)_5$ capsules, the Hb protein shells have a selective permeability. Similarly, the permeability decreases with the increase of wall thickness. Using FITC–dextran with a molecular weight of 20 kDa as a fluorescent probe, the fluorescence recovery in the capsule interior as a function of time is observed at lower excitation intensity. As a consequence, $(GA/Hb)_5$ microcapsules have a permeability of $4 \times 10^{-8}\,m\,s^{-1}$.

In a subsequent work, the adsorption of lipid bilayers onto protein multilayer capsules was performed [50–52]. As phosphatidylcholine lipid is a major component of biological membranes, egg phosphatidylcholine was chosen as a lipid model to cover the capsules, but a small fraction of negatively charged lipid, phosphatidic acid, was added in order to promote the adsorption and fusion of vesicles. In other words, the lipid composition on the surface could easily be varied and fine-tuned to specific conditions. The existence and stability of the lipid bilayer was proved by means of CLSM. The fluorescent orb arises from the presence and continuous distribution of lipids on the capsule shells.

In order to validate the use of lipid bilayer-coated protein multilayer capsules as a cell membrane model, an ion channel protein, H^+-ATP synthase, was incorporated in the supported lipid bilayer [46]. These supported membranes are prepared by lipid vesicle (or proteoliposome) fusion on protein multilayer capsules. The conservation of protein biological activity was checked by measuring the production of ATP. Moreover, for supported membranes containing active proteins, a biocompatible protein shell was inserted between the capsule and bilayer in order to separate proteins and support, and to favor their conformational integrity.

2.4
Application of Biomimetic Microcapsules

Selective delivery of therapeutic, diagnostic, and research compounds to specific sites or cells in biological environments could improve their efficacy and minimize potentially adverse side-effects. To this end, integrating specific biofunctionality on their carriers for targeting becomes promising in the field of medicine and drug delivery. Many biological or synthetic materials such as monoclonal antibodies, metabolites, peptide hormones, cytokines, growth factors, and viral and bacteriophage particles, could be employed as targeting entities. From the point of view of biomimetic design, a lipid bilayer coated on the LbL-assembled microcapsules provides the starting point for further integrating specific biofunctionality on the surface of biomimetic microcapsules for targeting.

2.4.1
Integrating Specific Biofunctionality for Targeting

It is well known that the surface of viruses carries proteins for recognition and interaction with the host cells so that they can easily pass across the barriers of their host cells and control the transcription/translation machinery of the cell. Viruses, therefore, have been successfully used as building blocks for the fabrication of composite functional materials so that selective delivery into cells and tissues can be achieved. Fischlechner *et al.* recently demonstrated that lipid bilayer-coated LbL-assembled microcapsules fused with viral particles at low pH, mimicking the conditions within the endosome of mammalian cells, can facilitate membrane passage [59–62]. In their first report, rubella-like particles (RLPs) were

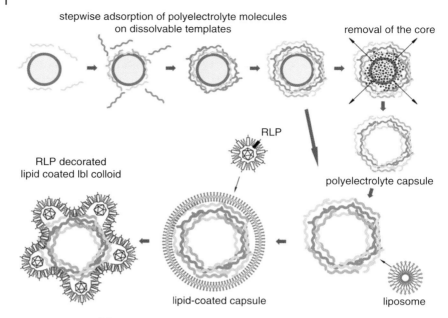

Figure 2.9 Protocol for engineering virus functionality on polyelectrolyte lipid composite capsules. (Reprinted with permission from [60]. © 2005, Wiley).

selected as a model virus (Figure 2.9) [59]. These particles are identical copies of the lipid-enveloped rubella virus except without the viral RNA. Rubella virus binds to its host cell surface at physiological pH, while fusion with the membrane takes place under acidic conditions. After incubation of PSS/PAH capsules with the positively charged PAH as the top layer and negatively charged phosphatidylserine vesicles, phosphatidylserine bilayer-coated PAH/PSS capsules were formed. At low pH (pH 4), the RLPs could attach to the phosphatidylserine layer by electrostatic attraction and subsequently fuse with the phosphatidylserine membrane. Tryptophan fluorescence spectroscopy and CLSM detected the presence of the RLPs when the virus was labeled with tetramethylrhodamine. The octadecylrhodamine (R18) dequenching assay found that at least a quarter of the applied RLPs had fused with the phosphatidylserine layer at pH 4, whereas fusion of not more than a few percent at neutral pH was observed. Low pH values are in favor of viral adsorption and can initiate subsequent fusion of the RLPs with the supported lipid layer. Fischlechner *et al.* ascribed this behavior to two factors. One is the virus surface becoming positively charged in acidic buffer, which has strong interaction with the negatively charged lipid membrane under these conditions. The other is that the pH changes cause the exposure of the hydrophobic peptides of the virus membrane fusion proteins, which can partially inert the hydrophobic core of the lipid bilayer.

In their subsequent studies, other virus particles such as influenza A/PR8 and baculovirus have also been used to demonstrate the general aspect of engineering virus functionalities on lipid-coated polyelectrolyte capsules for the design of bio/nonbio interfaces [60, 61]. All their studies show that the virus particles can firmly integrate in the supported membrane at low pH values and cannot be removed by washing, demonstrating the fusion of the virus particles with the lipid-coated polyelectrolyte capsules. Cell culture experiments show that these virus-engineered capsules were readily taken up into cells by endocytosis. Further research showed that the stability of polyelectrolyte capsules within a biological environment greatly influences the decomposition of the virus-like surface and subsequent propagation in the host cells. Additionally, the technological potential of these virus-decorated biomimetic capsules in diagnostic and other fields was also demonstrated through a bead assay method. Basically, the lipid membrane on the surface of LbL-assembled capsules has at least two key functions – it serve as a means for virus fusion during the process of fabrication and inhibits nonspecific interactions with the antibodies. The latter is a major problem associated with surface-bound antigens.

2.4.2
Adsorption of Antibodies on the Surface of Biomimetic Microcapsules

In immunotherapeutic approaches to cancer therapy, a strategy to achieve targeting utilizes antibodies that bind specifically to antigens on the surface of tumor cells. This strategy can similarly be used to realize the biofunctionalization of biomimetic microcapsules. Caruso *et al.* recently demonstrated this possibility of targeting and uptake of specific antibody-modified LbL-assembled capsules [63]. First, a dilauroylphosphatidylethanolamine (DLPE) bilayer was assembled on the outer surface of polyelectrolyte capsules. The monoclonal immunoglobulin G (IgG) antibodies (i.e., a primary antibody, mouse IgG) were then attached onto the DLPE-modified capsules via noncovalent interaction to impart biospecificity. After they were exposed to a FITC-labeled secondary antibody (rabbit antimouse IgG) solution, homogeneous surface coverage of the FITC-labeled secondary antibody was observed by fluorescence microscopy, confirming that the primary antibody was homogeneously attached to the DLPE coating.

In a subsequent study, Caruso *et al.* immobilized the humanized A33 monoclonal antibody, which can bind to a transmembrane glycoprotein, the human A33 antigen, onto the LbL-assembled capsules via noncovalent interaction [64]. The human A33 antigen is expressed by 95% of human colorectal tumor cells as well as on the basolateral surfaces of intestinal epithelial cells, but not by other epithelial tissues. The cell culture experiments confirmed that the humanized A33 monoclonal antibody-decorated particles can be targeted and selectively internalized by tumor cells, while showing minimal affinity for cells that do not express the A33 antigen, confirming that the selective binding was performed through antibody–antigen specific recognition.

2.5
Conclusions and Perspectives

As a powerful molecular assembly technique, LbL assembly has been extensively used to fabricate hollow multifunctionalized capsules due to their potential application in the fields of materials science, biology, and medicine. In this chapter, we have outlined the latest progress of biomimetic microcapsule fabrication by LbL assembly including mimicking the functionalities of cells such as selective ion transportation, integration of biological components onto the microcapsules, and improvement of the fabrication efficiency. One important application of these biomimetic microcapsules is to serve as cellular mimics to regenerate some cellular processes in a man-made environment. The other important application of these biomimetic microcapsules is the development of novel drug delivery systems. An important issue in this potential application is how to promote their cellular uptake and how to target them to cells. So far, biofunctionalization is a good choice because it can mimic biological behavior. For instance, lipid bilayer-coated polyelectrolyte shells, by incorporating a variety of virus-like particles or by coupling antibodies, are promising with regard to the selective uptake of biomimetic capsules by specific cell types.

References

1 Ratner, B.D. and Bryant, S.J. (2004) Biomaterials: where we have been and where we are going. *Annu. Rev. Biomed. Eng.*, **6**, 41–75.
2 Sanchez, C., Arribart, H., Madeleine, M., and Guille, G. (2005) Biomimetism and bioinspiration as tools for the design of innovative materials and systems. *Nat. Mater.*, **4**, 277–288.
3 Heuer, A.H., Fink, D.J., Laraia, V.J., Arias, J.L., Calvert, P.D., Kendall, K., Messing, G.L., Blackwell, J., Rieke, P.C., Thompson, D.H., Wheeler, A.P., Veis, A., and Caplan, A.I. (1992) Innovative materials processing strategies: a biomimetic approach. *Science*, **255**, 1098–1105.
4 Sarikaya, M., Tamerler, C., Yen, A.K.-Y., Schulten, K., and Baneyx, F. (2003) Molecular biomimetics: nanotechnology through biology. *Nat. Mater.*, **2**, 577–585.
5 He, Q., Cui, Y., and Li, J.B. (2009) Molecular assembly and application of biomimetic microcapsules. *Chem. Soc. Rev.*, **38**, 2292–2303.
6 He, Q., Duan, L., Qi, W., Wang, K.W., Cui, Y., Yan, X.H., and Li, J.B. (2008) Microcapsules containing a biomolecular motor for ATP biosynthesis. *Adv. Mater.*, **20**, 2933–2937.
7 Whitesides, G.M., Mathias, J.P., and Seto, C.T. (1991) Molecular self-assembly and nanochemistry: a chemical strategy for the synthesis of nanostructures. *Science*, **254**, 1312–1319.
8 Bao, G. and Suresh, S. (2003) Cell and molecular mechanics of biological materials. *Nat. Mater.*, **2**, 715–725.
9 Mitragotri, S. and Laham, J. (2009) Physical approaches to biomaterial design. *Nat. Mater.*, **8**, 15–23.
10 Lasic, D.D. (1993) *Liposomes: From Physics to Applications*, Elsevier, Amsterdam.
11 Barenholz, Y. (2001) Liposome applications: problems and prospects. *Curr. Opin. Colloid Interface Sci.*, **6**, 66–77.
12 Wendell, D., Patti, J., and Montemagno, C.D. (2006) Using biological inspiration to engineer functional nanostructured materials. *Small*, **2**, 1324–1329.

13 Sukhorukov, G.B. and Möhwald, H. (2006) Multifunctional cargo systems for biotechnology. *Trends Biotechnol.*, **25**, 93–98.
14 Decher, G. and Hong, J.D. (1991) Buildup of ultrathin multilayer films by a self-assembly process: consecutive adsorption of anionic and cationic bipolar amphiphiles on charged surfaces. *Macromol. Chem. Macromol. Symp.*, **46**, 321–327.
15 Decher, G. (1997) Fuzzy nanoassemblies: toward layered polymeric multicomposites. *Science*, **277**, 1232–1237.
16 Li, J.B., Möhwald, H., An, Z.H., and Lu, G. (2005) Molecular assembly of biomimetic microcapsules. *Soft Matter*, **1**, 259–264.
17 Peyratout, C.S. and Daehne, L. (2004) Tailor-made polyelectrolyte microcapsules: from multilayers to smart containers. *Angew. Chem. Int. Ed.*, **43**, 3762–3783.
18 Quinn, J.F., Johnston, A.P.R., Such, G.K., Zelikin, A.N., and Caruso, F. (2007) Next generation, sequentially assembled ultrathin films: beyond electrostatics. *Chem. Soc. Rev.*, **36**, 707–718.
19 He, Q., Cui, Y., Ai, S.F., Tian, Y., and Li, J.B. (2009) Self-assembly of composite nanotubes and their applications. *Curr. Opin. Colloid Interface Sci.*, **14**, 115–125.
20 Donath, E., Sukhorukov, G.B., Caruso, F., Davis, S.A., and Möhwald, H. (1998) Novel hollow polymer shells by colloid-templated assembly of polyelectrolytes. *Angew. Chem. Int. Ed.*, **37**, 2201–2205.
21 Sukhorukov, G.B., Donath, E., Davis, S., Lichtenfeld, H., Caruso, F., Popov, V.I., and Moehwald, H. (1998) Stepwise polyelectrolyte assembly on particle surfaces: a novel approach to colloid design. *Polym. Adv. Technol.*, **9**, 759–767.
22 Voigt, A., Lichtenfeld, H., Zastrow, H., Sukhorukov, G.B., Donath, E., and Moehwald, H. (1999) Membrane filtration for microencapsulation and microcapsules fabrication by layer-by-layer polyelectrolyte adsorption. *Ind. Eng. Chem. Res.*, **38**, 4037–4043.
23 Sukhorukov, G.B., Brumen, M., Donath, E., and Moehwald, H. (1999) Hollow polyelectrolyte shells: exclusion of polymers and Donnan equilibrium. *J. Phys. Chem. B*, **103**, 6434–6440.
24 G.Ibarz, L.D., Donath, E., and Moehwald, H. (2002) Resealing of polyelectrolyte capsules after core removal. *Macromol. Rapid Commun.*, **23**, 474–478.
25 Ariga, K., Hill, J.P., and Ji, Q.M. (2007) Layer-by-layer assembly as a versatile bottom-up nanofabrication technique for exploratory research and realistic application. *Phys. Chem. Chem. Phys.*, **9**, 2319–2340.
26 Johnston, A.P.R., Cortez, C., Angelatos, A.S., and Caruso, F. (2006) Layer-by-layer engineered capsules and their applications. *Curr. Opin. Colloid Interface Sci.*, **11**, 203–209.
27 Volodkin, D.V., Petrov, A.I., Prevot, M., and Sukhorukov, G.B. (2004) Matrix polyelectrolyte microcapsules: new system for macromolecule encapsulation. *Langmuir*, **20**, 3398–3406.
28 Lvov, Y., Antipov, A.A., Mamedov, A., Moehwald, H., and Sukhorukov, G.B. (2001) Urease encapsulation in nanoorganized microshells. *Nano Lett.*, **1**, 125–128.
29 Tiourina, O.P., and Sukhorukov, G.B. (2002) Multilayer alginate/protamine microsized capsules: encapsulation of α-chymotrypsin and controlled release study. *Int. J. Pharm.*, **242**, 155–161.
30 Wang, Y.J., Yu, A.M., and Caruso, F. (2005) Nanoporous polyelectrolyte spheres prepared by sequentially coating sacrificial mesoporous silica spheres. *Angew. Chem. Int. Ed.*, **44**, 2888–2892.
31 Wang, Y.J., Angelatos, A.S., Dunstan, D.E., and Caruso, F. (2007) Infiltration of macromolecules into nanoporous silica particles. *Macromolecules*, **40**, 7594–7600.
32 Tong, W.J. and Gao, C.Y. (2008) Multilayer microcapsules with tailored structures for bio-related applications. *J. Mater. Chem.*, **18**, 3799–3812.
33 Tong, W.J., Gao, C.Y., and Moehwald, H. (2005) Manipulating the properties of polyelectrolyte microcapsules by glutaraldehyde cross-linking. *Chem. Mater.*, **17**, 4610–4616.
34 Grigoriev, D.O., Bukreeva, T., Moehwald, H., and Shchukin, D.G. (2008) New method for fabrication of loaded

micro- and nanocontainers: emulsion encapsulation by polyelectrolyte layer-by-layer deposition on the liquid core. *Langmuir*, **24**, 999–1004.

35 Teng, X.R., Shchukin, D.G., and Moehwald, H. (2008) A novel drug carrier: lipophilic drug-loaded polyglutamate/polyelectrolyte nanocontainers. *Langmuir*, **24**, 383–389.

36 Antipov, A.A. and Sukhorukov, G.B. (2004) Polyelectrolyte multilayer capsules as vehicles with tunable permeability. *Adv. Colloid Interface Sci.*, **111**, 49–61.

37 Glinel, K., Sukhorukov, G.B., Moehwald, H., Khrenov, V., and Tauer, K. (2003) Thermosensitive hollow capsules based on thermoresponsive polyelectrolytes. *Macromol. Chem. Phys.*, **204**, 1784–1790.

38 Tao, X., Moehwald, H., and Li, J.B. (2004) Self-assembly, optical behavior, and permeability of a novel capsule based on an azo dye and polyelectrolytes. *Chem. Eur. J.*, **10**, 3397–3403.

39 Qi, W., Yan, X., Duan, L., Cui, Y., Yang, Y., and Li, J.B. (2009) Glucose-sensitive microcapsules from glutaraldehyde cross-linked hemoglobin and glucose oxidase. *Biomacromolecules*, **10**, 1212–1216.

40 Qi, W., Yan, X.H., Fei, J.B., Wang, A.H., Cui, Y., and Li, J.B. (2009) Triggered release of insulin from glucose-sensitive enzyme multilayer shells. *Biomaterials*, **30**, 2799–2806.

41 Gao, C.Y., Leporatti, S., Moya, S., Donath, E., and Moehwald, H. (2001) Stability and mechanical properties of polyelectrolyte capsules obtained by stepwise assembly of poly(styrenesulfonate sodium salt) and poly(diallyldimethyl ammonium) chloride onto melamine resin particles. *Langmuir*, **17**, 3491–3495.

42 Ge, L.Q., Möhwald, H., and Li, J.B. (2003) Phospholipase A_2 hydrolysis of mixed phospholipid vesicles formed on polyelectrolyte hollow capsules. *Chem. Eur. J.*, **9**, 2589–2594.

43 Ge, L.Q., Möhwald, H., and Li, J.B. (2003) Biointerfacing polyelectrolyte microcapsules. *ChemPhysChem*, **4**, 1351–1355.

44 Ge, L.Q., Li, J.B., and Möhwald, H. (2003) Polymer-stabilized phospholipid vesicles formed on polyelectrolyte multilayer capsules. *Biochem. Biophys. Res. Comm.*, **303**, 653–659.

45 Duan, L., He, Q., Wang, K., Yan, X., Cui, Y., Möhwald, H., and Li, J.B. (2007) Adenosine triphosphate biosynthesis catalyzed by F_oF_1 ATP synthase assembled in polymer microcapsules. *Angew. Chem. Int. Ed.*, **46**, 6996–7000.

46 Qi, W., Duan, L., Wang, K., Yan, X., Cui, Y., He, Q., and Li, J.B. (2008) Motor protein CF_oF_1 reconstituted in lipid-coated hemoglobin microcapsules for ATP synthesis. *Adv. Mater.*, **20**, 601–605.

47 An, Z., Möhwald, H., and Li, J.B. (2006) pH controlled permeability of lipid/protein biomimetic microcapsules. *Biomacromolecules*, **7**, 580–585.

48 Li, J.B., Zhang, Y., and Yan, L. (2001) Multilayer formation on a curved drop surface. *Angew. Chem. Int. Ed.*, **40**, 891–894.

49 He, Q., Zhang, Y., Lu, G., Miller, R., Möhwald, H., and Li, J.B. (2008) Dynamic adsorption and characterization of phospholipid and mixed phospholipid/protein layers at liquid/liquid interfaces. *Adv. Colloid Interface Sci.*, **140**, 67–76.

50 An, Z.H., Tao, C., Lu, G., Möhwald, H., Zheng, S.P., Cui, Y., and Li, J.B. (2005) Fabrication and characterization of human serum albumin and L-alpha-dimyristoylphosphatidic acid microcapsules based on template technique. *Chem. Mater.*, **17**, 2514–2519.

51 An, Z.H., Lu, G., Möhwald, H., and Li, J.B. (2004) Self-assembly of human serum albumin (HSA) and L-alpha-dimyristoylphosphatidic acid (DMPA) microcapsules for controlled drug release. *Chem. Eur. J.*, **10**, 5848–5852.

52 Duan, L., Qi, W., Yan, X.H., He, Q., Cui, Y., Wang, K., Li, D.X., and Li, J.B. (2009) Proton gradients produced by glucose oxidase microcapsules containing motor F_oF_1-ATPase for continuous ATP biosynthesis. *J. Phys. Chem. B*, **113**, 395–399.

53 Moya, S., Donath, E., Sukhorukov, G.B., Auch, M., Baeumler, H., Lichtenfeld, H., and Möhwald, H. (2000) Lipid coating on polyelectrolyte surface modified colloidal particles and polyelectrolyte capsules. *Macromolecules*, **33**, 4538–4544.

54 Moya, S., Richter, W., Leporatti, S., Baeumler, H., and Donath, E. (2003) Freeze-fracture electron microscopy of lipid membranes on colloidal polyelectrolyte multilayer coated supports. *Biomacromolecules*, **4**, 808–814.

55 He, Q. and Li, J.B. (2007) Hydrolysis characterization of phospholipid monolayers catalyzed by different phospholipases at the air–water interface. *Adv. Colloid Interface Sci.*, **131**, 91–98.

56 Katagiri, K. and Caruso, F. (2005) Monodisperse polyelectrolyte-supported asymmetric lipid-bilayer vesicles. *Adv. Mater.*, **17**, 738–743.

57 He, Q., Tian, Y., Möhwald, H., and Li, J.B. (2009) Biointerfacing luminescent nanotubes. *Soft Matter*, **5**, 300–303.

58 Duan, L., He, Q., Yan, X.H., Cui, Y., Wang, K.W., and Li, J.B. (2007) Hemoglobin protein hollow shells fabricated through covalent layer-by-layer technique. *Biochem. Biophys. Res. Comm.*, **354**, 357–362.

59 Fischlechner, M., Zaulig, M., Meyer, S., Estrela-Lopis, I., Cuellar, L., Irigoyen, J., Pescador, P., Brumen, M., Messner, P., Moya, S., and Donath, E. (2008) Lipid layers on polyelectrolyte multilayer supports. *Soft Matter*, **4**, 2245–2258.

60 Fischlechner, M., Zchoernig, O., Hofmann, J., and Donath, E. (2005) Engineering virus functionalities on colloidal polyelectrolyte lipid composites. *Angew. Chem. Int. Ed.*, **44**, 2892–2895.

61 Fischlechner, M., Reibetanz, U., Zaulig, M., Enderlein, D., Romanova, J., Leporatti, S., Moya, S., and Donath, E. (2007) Fusion of enveloped virus nanoparticles with polyelectrolyte-supported lipid membranes for the design of bio/nonbio interfaces. *Nano Lett.*, **7**, 3540–3546.

62 Fischlechner, M., Toellner, L., Messner, P., Grabherr, R., and Donath, E. (2006) Virus-engineered colloidal particles – a surface display system. *Angew. Chem. Int. Ed.*, **45**, 784–789.

63 Angelatos, A.S., Radt, B., and Caruso, F. (2005) Light-responsive polyelectrolyte/gold nanoparticle microcapsules. *J. Phys. Chem. B*, **109**, 3071–3076.

64 Cortez, C., Tomaskovic-Crook, E., Johnston, A.P.R., Radt, B., Cody, S.H., Scott, A.M., Nice, E.C., Heath, J.K., and Caruso, F. (2006) Targeting and uptake of multilayered particles to colorectal cancer cells. *Adv. Mater.*, **18**, 1998–2003.

3
F$_o$F$_1$-ATP Synthase-Based Active Biomimetic Systems

3.1
Introduction

Biomimetics has proved very useful in the design and fabrication of new functional structured materials on the micro- and nanoscale. Biomimetics refer to human-made processes, devices, or systems that mimic or imitate certain aspects of biological systems and have proven useful in providing biological inspiration from natural efficient designs [1–4]. Engineering biomimetic materials encompasses a wide variety of research, from current nanomaterials such as self-cleaning glass and artificial shark skin, to the mechanics of how biological molecules such as proteins, enzymes, DNA, and RNA can function as analogous man-made structures. In fact, biomimetics is not limited to just copying nature because, with the development of modern biology, scientists can directly utilize biological units themselves to construct hybrid nanostructured materials. Thus, some of the manufacturing difficulties of biomimetics can be avoided. Active proteins such as kinesin, myosin, ATP synthase (or ATPase) are called "molecular motors," and play essential roles in the activities of cells and regulate specific functions through their stimuli-responsive mechanical motions [5, 6]. The confluence of scientific developments in modern biology and nanoscience now offers the potential to design functional hybrid nanomaterials. A major challenge for the construction of functional hybrid nanomaterials is how to integrate natural molecular machines such as motor proteins into the engineering of active biomimetic systems. In this chapter, we explore how biomimetics applied to engineering functional nanomaterials, particularly assembling ATPase in artificial containers and mimetic cellular systems with cellular processes.

3.2
F$_o$F$_1$-ATPase – A Rotary Molecular Motor

Molecular motors such as myosin, kinesin, dynein, and ATPase abound in the living cells. Myosin motors drive muscle contraction and kinesin or dynein motors

Molecular Assembly of Biomimetic Systems. Junbai Li, Qiang He, and Xuehai Yan
© 2011 WILEY-VCH Verlag GmbH & Co. KGaA, Weinheim
ISBN: 978-3-527-32542-9

transport vesicles from one end of the cell to the other. All of these linear motors are supplied by the biological energy through the hydrolysis of ATP molecules. The F_oF_1-ATPase is responsible for the catalytic synthesis of ATP molecules in biological organisms. They are widely present in the membranes of mitochondria, chloroplasts, and prokaryotic cells, where they convert transmembrane electrochemical proton gradients into ADP~P bonds (i.e., ATP). This reaction is commonly called ATP synthesis. ATP production is one of the major chemical reactions in living organisms. It has been estimated that a human uses 40 kg of ATP in normal daily living. On the other hand, the reverse reaction, the hydrolysis of ATP, is coupled with a very great many biochemical processes and supplies the energy needed for these processes. ATP can thus be considered as the cell's "fuel" supply, consumed all the time and constantly recycled. In ATPase, the ATP synthesis reaction is coupled to an exergonic flow of H^+ ions resulting from a transmembrane electrochemical potential difference, itself generated by redox reactions in the membrane (respiratory or photosynthetic system). In the following section, structural and mechanical aspects of $H^+F_oF_1$-ATPase are briefly described since $H^+F_oF_1$-ATPase is the best known example of this family.

3.2.1
Structure of $H^+F_oF_1$-ATPase

The structure of $H^+F_oF_1$-ATPase in different functional forms is now being rapidly elucidated. It has been confirmed that this enzyme is comprised of two separate domains: F_o, the hydrophobic membrane-bound portion that is responsible for proton translocation, and F_1, which is responsible for ATP hydrolysis or synthesis. The extramembranous F_1 catalytic subcomplex is attached to the membrane intrinsic F_o subcomplex by a central stalk and a peripheral stalk. Synthesis of ATP in F_1 is driven by rotation of the central stalk, driven by protons passing through the membrane domain F_o [6]. The enzyme can also catalyze the reverse reaction, with ATP hydrolysis in F_1 driving proton pumping in F_o. Figure 3.1 summarizes the structure of F_oF_1-ATPase. The protein has a bipartite structure as its name suggests, with the membrane-intrinsic portion, F_o, responsible for proton transport and F_1 performing the task of ATP synthesis/hydrolysis. When F_1 is detached from F_o and is therefore decoupled from the proton-motive driving force, it catalyzes only ATP hydrolysis. Depending on the organism, this enzyme consists of at least eight different subunits and more than 20 polypeptide chains. Generally, there is an F_1 part of α, β, γ, δ, and ε subunits, and an F_o part, comprising a, b and c subunits in the stoichiometry $1:2:10-14$. F_1 and F_o are linked together by two stalks – a central one containing the γ and ε subunits, and a peripheral one involving the δ and b subunits. The rotor is formed by the central stalk and a ring of c subunits in F_o that contain an essential carboxylic group. Rotation is generated by the binding and release of protons to the carboxylic group in the c ring as it enters the interface with the membrane subunit a of F_o [8–10]. The peripheral stalk acts as a stator that prevents the $\alpha\beta$ subcomplex in F_1 from following the rotation of the central stalk. In other words, ATPase is a combination of two motors, termed

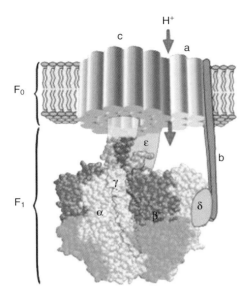

Figure 3.1 Arrangement of the subunits in the F_oF_1-ATPase complex. One α subunit has been removed from the F_1 part to reveal the γ subunit with the αβ domain. There are three αβ subunit pairs. The αβ domain is attached to the α subunit in the F_o part by a peripheral stalk composed of δ and two copies of the b subunit. The c subunit ring of F_o is linked to the γ and ε subunits to form the central rotor. (Reprinted with permission from [7]. © 1998, Nature Publishing Group.)

F_o and F_1. The F_o motor is powered by a proton electrochemical potential and the F_1 motor is powered by ATP. In the complex, the two motors are connected so that it acts as both a motor and a generator. Motion of F_o driven by proton flow can be used to generate ATP in F_1. Alternatively, by rotating in the opposite direction, ATP-powered motion of F_1 can be used to turn F_o and pump protons, creating an electrochemical gradient.

The F_o motor, driven by protons, is connected to the F_1 motor through an axle. The F_1 motor is driven by the cleavage of ATP. F_1 is composed of a ring of six subunits, three α subunits. The different conformations of the β subunit have markedly different affinity for ATP. Boyer first identified three forms, labeled tight, loose, and open, and more details were added later [7, 11–13]. In order to explain the unusual properties between F_o and F_1 parts, he then proposed what has become known as the "binding change" or "alternating site" hypothesis. The key feature of this hypothesis is that the three catalytic sites, and therefore the three αβ subunit pairs containing these sites, are independent in different conformations at any one time [14, 15]. One is open and ready for ATP (or ADP + P_i) binding, while the second and third are partly open and closed, respectively, around bound nucleotide. ATP binding, and the resulting closure of the open site, produces a cooperative conformational change in which the other two sites are altered, so that

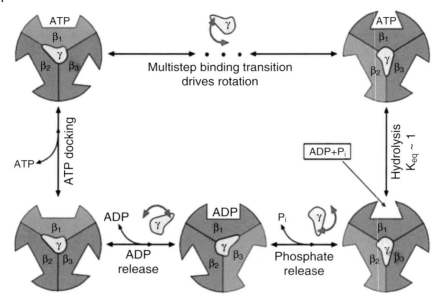

Figure 3.2 Bing mechanism model of ATPase forms assumed by one catalytic site of the ATPase during synthesis or hydrolysis of ATP. (Reprinted with permission from [17]. © 2000, Elsevier.)

the closed one becomes partly open and the partly open one becomes fully open. Thus, each site alternates between the three states as ATP hydrolysis or, in the reverse direction, as ATP synthesis, proceeds.

The above-mentioned binding change mechanism gives rise to two obvious questions – how are the conformational changes propagated between the three catalytic sites in F_1F_o and how are the catalytic site events are coupled to proton translocation? An idea suggested by Boyer proposed that rotation of one or more of the single copy subunits in F_1 could occur [12, 16]. Later, it was recognized as the γ and ε subunits providing the rotor as shown in Figure 3.2 [16, 17]. At the same time, Cox *et al.* independently developed an idea that F_1F_o works as a rotary motor, but with a focus on the proton translocation mechanism in the F_o part [18–21]. All these studies insightfully proposed that the c subunit ring rotates against the a and b subunits during proton translocation. Recent experiments applying diverse innovative biochemical and physical approaches have successfully confirmed this hypothesis. These experimental findings greatly helped us to understand the motor mechanism of the ATPase [22–24]. For example, the F_1 part of the *Escherichia coli* complex was visualized by using cryoelectron microscopy through the top or bottom [25, 26]. Owing to the intrinsic asymmetry of the molecule, details of the subunit arrangement were resolved. The results showed that the α and β subunits alternate around a hexagon containing a central mass that could be identified as the γ subunit. When the individual images were classified on the basis of their dominant features, they fell into three classes, with the γ subunit being located at a different αβ pair in each. High-resolution X-ray analysis confirmed basic tenets of the alternating sites hypothesis [27]. It showed an enzyme in which the three catalytic sites had different conformations: one open, one closed for ongoing bond cleavage, and the third partly open for imminent release of product ADP and P_i. It had been shown that cross-linking of γ to the C-terminal domain of β blocked activity. Based on this, Capaldi *et al.* showed that, if the cross-link was subsequently released and reformed, γ became attached to a different β subunit [28]. Junge *et al.* established a rotation of γ through at least 280° by fluorescently labeling the C-terminal amino acid of this subunit in chloroplast F_1 and then measuring ATP hydrolysis-driven rotation by polarized absorption recovery after photobleaching [29].

3.2.2
Direct Observation of the Rotation of Single ATPase Molecules

Direct visualization of the rotation of the c subunits against the a subunit is necessary to dissect out the kinetics and stepping as a function of enzyme turnover. Stepping here refers to the number of discrete rotational steps of the c subunit ring against the a subunits for each 120° rotation of the γε subunit pair with respect to the αβ pair in F_1. However, demonstrating rotation of the c subunit ring in the intact F_1F_o complex is proving problematic. Several groups have reported visualizing ATP-driven rotation of the c subunit ring when monitored by the fluorescent actin filament method. In 1997, Noji *et al.* first succeeded in direct observation of the rotation of the F_1 motor by means of a microprobe technique [29]. Briefly, a biotin unit was attached to the F_1 motor through a cysteine residue, which had been introduced by site-directed mutagenesis, and the resulting conjugate was connected to a histidine decamer at the N-terminus of each β subunit. This engineered F_1 motor was immobilized on a glass plate, whose surface had been covered by Ni^{2+}-nitrilotriacetic acid (Ni-NTA) with a high affinity toward histidine oligomers. Then, a fluorescently labeled biotinylated actin filament was grafted via streptavidin to the base of the γ subunit on the extrinsic part of a bacterial F_1-ATPase, the upper part of the $\alpha_3\beta_3$ crown being bound onto a substrate (Figure 3.3a). Rotation of the actin filament induced by ATP hydrolysis was then observed by fluorescence microscopy. This group saw that ATP hydrolysis drove a 360° rotation of the actin filament in three 120° steps in one direction (counterclockwise when viewing the enzyme from below) with very few reversals as shown in Figure 3.3(b). (Videos of this dramatic experiment can be seen at http://www.res.titech.ac.jp/seibutu/main_.html.) Owing to the solvent viscous drag on the actin filament, the rate of rotation was very slow – only 3% of the enzyme turnover rate. This problem has been largely overcome in recent experiments by tagging the γ subunit with much smaller (40 nm) gold particles (Figure 3.4) [30]. In their study, Kinosita, Yoshida and colleagues observed rotation rates of 134 revolutions s^{-1}, which is the rate expected at steady-state ATP hydrolysis under the conditions used. The researchers were then able to dissect out two substeps in the rotation of γ between

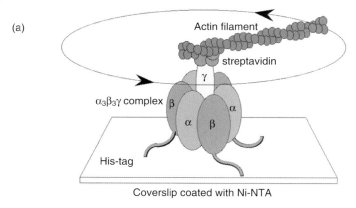

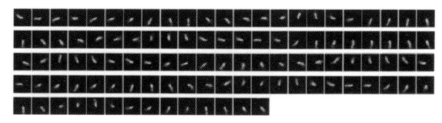

Figure 3.3 (a) First single-molecule experiments on ATPase. The enzyme was bound to a substrate by its stator part (β subunits) and a fluorescent actin filament grafted onto the rotor (γ or ε subunit for the F_1 subcomplex, c subunits for the F_oF_1 complex). The substrate was a glass slide coated with a nickel-charged resin with a strong affinity for histidine. The enzyme anchoring was achieved by a polyhistidine (His) tag introduced by mutagenesis at the N-terminus of the β subunits. By introducing a cysteine onto the rotor, it could be biotinylated and hence bound via a streptavidin molecule to the actin filament, itself biotinylated. (b) In the presence of ATP, its rotation could be observed in a small proportion of the molecules (about 1%) by fluorescence microscopy. Seen from above, it occurs in the anticlockwise direction. Successive views of the actin filament are shown schematically on the right. (Reprinted with permission from [30]. © 1997, Nature Publishing Group.)

two αβ pairs—one of around 90°, which they attribute to ATP binding, and a second of around 30°, which occurs with product release. These two substeps were separated by a short dwell time. This experiment was much refined in later experiments [31–34]. The experiments originally carried out on the isolated F_1 part were extended to the whole ATPase, no longer observing the rotation of the central stalk of the F_1 part, but instead of the ring of c subunits that form the membrane part of the rotor.

Another important work performed by Montemagno et al. showed an ATP-powered motor with nanofabricated propellers linked to genetically engineered protein F_1-ATPase [35–37]. They fabricated arrayed F_1 motors on an engineered

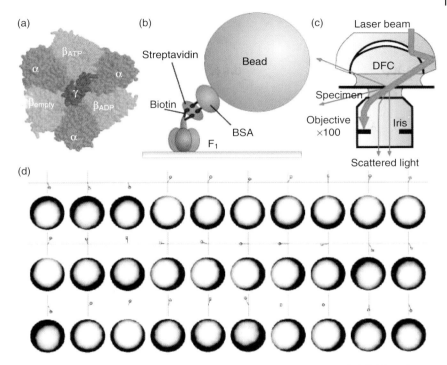

Figure 3.4 Observation of F_1 rotation. (a) Atomic structure of F_1-ATPase viewed from the F_o side (top in b). (b) Side view of the observation system. The 40-nm bead gave a large enough optical signal that warranted a submillisecond resolution, but the bead was small enough not to impede the rotation. (c) Laser dark-field microscopy for observation of gold beads. Only light scattered by the beads exited the objective and was detected. DFC, dark-field condenser. (d) Sequential images of a rotating bead at 2 mM ATP. Images are trimmed in circles (diameter 370 nm) to aid identification of the bead position; centroid positions are shown above the images at ×3 magnification. The interval between images is 0.5 ms (Reprinted with permission from [31]. © 2001, Nature Publishing Group.)

surface bearing nanoscale nickel posts, and they have succeeded in rotating the nickel-based nanopropellers attached to the c subunit of F_1 motors. The rotation of the nanopropellers is initiated by 2 M ATP, while it is inhibited by sodium azide. When a F_1 motor having metal-binding sites is used, the rotary motion can be switched on and off reversibly in response to a change in [Zn^{2+}]. Rotary motions of the F_1 motor driven by external forces have been utilized for the synthesis of ATP. For example, Dimroth *et al.* have reported the synthesis of ATP by a voltage-driven rotation of the F_o motor using an electrical potential generated with a proteoliposome membrane [38]. Hisabori *et al.* have incorporated a redox-active functionality into a bacterial F_1 motor to control its rotary motion by a redox reaction [39]. Recently, Itoh *et al.* have reported the fabrication of an engineered F_1 motor by attaching a magnetic bead to its c subunit and they have demonstrated the synthesis of ATP by rotating the bead in an appropriate direction using electromagnets [40].

The rotational motion of F_1-ATPase was coupled with hydrolysis of ATP. The work showed that F_1-ATPase therefore possesses energy transduction properties (i.e., can convert chemical energy to mechanical energy). As a rotary molecular motor and ATP producer, the ATPase has attracted great interest and demonstrated numerous potential applications, from the generation of bioenergy to the design of novel nanodevices. Overall, direct proof of the rotation of the central stalk within F_1-ATPase hydrolysis was provided by monitoring a fluorescently labeled actin filament or a micrometer particle to the γ subunit of the F1-ATPase. This experiment provided important insights into the mechanism of this molecular motor.

3.3
Reconstitution of F_oF_1-ATPase in Cellular Mimic Structures

In natural biological systems, the plasma membrane formed by the self-assembly of lipid molecules is a flexible barrier that serves to separate the cell's internal environment from its external environment. This biological membrane is composed of a closed lipid bilayer structure embedded with various membrane proteins. In the following section, three biomimetic membrane systems including liposomes, lipid-modified polymer multilayer capsules, and polymersomes will be discussed.

In the past, lipid membranes have served as useful platforms for *in vitro* protein reconstitution in basic structural and mechanistic studies of membrane proteins. Among of them, the liposome is the most popular biomembrane model because compartmentalization is a fundamental requirement for the reproduction of the natural environment of membrane-bound proteins. Thus, a protein-incorporated liposome (proteoliposome) can be regarded as a type of functionalized compartment [41]. However, as stated in Chapter 2, limitations of size, and mechanical and chemical stability of the assembled liposomes actually result in difficulties in application. To address this issue, supported membrane systems have been produced [42]. In Chapter 2, we showed that lipid-coated polyelectrolyte microcapsules could be fabricated through the conversion of liposomes into lipid bilayers to cover the capsules' surface in analogy to the cell membrane. The stability and lifetime of lipid membranes have been greatly improved due to the support of polyelectrolyte multilayer shells. Additionally, polymer membranes could offer a compelling approach towards fabricating customizable, rugged devices because certain parameters such as durability and chain length could easily be tuned. A triblock copolymer, with hydrophilic ends and a hydrophobic core, can provide an environment very similar to a lipid. In this case, the lipid membranes can be replaced with polymer. Thus, both lipid-modified polymer multilayer microcapsules and polymersomes can be considered as ideally supported biomimetic membrane systems to mimic real cell membrane structures. At the same time, these new hybrid systems also enable the design and application of new biomimetic structured materials.

3.3.1
F_oF_1-ATPase-incorporated Liposome – A Classical Biomembrane Mimic

In most cases, studies of membrane proteins in their native environment can be difficult to interpret due to restrictions arising from the complexity of the native membranes and interference with other membrane constituents or other reactions. Since the pioneering work of Racker *et al.* in the 1970s, liposomes incorporating purified membrane proteins (i.e., proteoliposomes) have become a powerful tool for elucidating both the functional and structural aspects of the membrane-associated proteins [41, 43–45]. This approach has been applied with success to a very diverse range of membrane functions. Within this framework, membrane reconstitution has played a central role in the studies of membrane-bound proteins with a vectorial function (transport) and, in particular, has helped to improve our understanding of the mode of operation of energy-transducing enzymes such as ATPases. Reconstitution has made an especially important contribution to the studies dealing with the mechanisms of transport, the determination of nature of the transported ions, the electrical properties of the transport mechanism, and the analysis of the coupling between transport systems. Of course, the concepts are not restricted to transport proteins, since reconstitution also allows analysis of other important general properties of membrane-bound proteins such as lipid–protein and protein–protein interactions, the topological features of proteins in the membrane, the specific roles of different subunits or ligand recognition and binding [45]. However, in this section, we only focus on the reconstitution of ATPases in the liposomes (a number of reviews or book chapters are available concerning specific classes of membrane proteins such as receptors, substrate carriers, and energy-conserving enzymes involved in oxidative phosphorylation). Concerning the reconstitution of membrane-bound proteins into liposomes, four main technical strategies have been developed: organic solvent-mediated reconstitution, mechanical means, direct incorporation into preformed liposomes, and detergent-mediated reconstitution [41, 45]. Among them, detergent-mediated reconstitution is the most successful and frequently used strategy for proteoliposome preparation because most membrane-bound proteins are isolated and purified in the presence of detergents. In a general procedure, the membrane-bound proteins are first cosolubilized with phospholipids in the appropriate detergent in order to form an isotropic solution of lipid–protein–detergent and lipid–detergent micelles. Next, the detergent is removed, resulting in the progressive formation of bilayer vesicles with incorporated protein. In the following, all proteoliposomes are prepared by applying this approach.

3.3.1.1 Bacteriorhodopsin uses Light to Pump Protons
As discussed in Section 3.2.1, electrochemical gradients can be used to power diverse biomolecular processes. The electron transport chain created by proteins

in cells uses the flow of electrons to power the transport of electrons across the membrane. These proteins contain a string of electron-carrying cofactors, ranging from weak carriers to carriers with strong affinity for electrons. As the electrons travel from unstable to stable carriers, the energy of the flow is used to translocate protons across the membrane. Two mechanisms are proposed [46]. The first mechanism is that they may guide the docking of a proton-carrying cofactor, orienting it first on one side of the membrane to pick up protons, then moving it to the other side to release them. The second mechanism is that the flow of electrons may force allosteric changes in the protein structure similar to the ABC transporters, opening gates on one side and then the other.

Bacteriorhodopsin is a membrane-bound protein originally isolated from the purple membranes of *Halobacterium halobium* and also is one of the best-understood proton pumps. Powered by light, the protein cycles through three states of different energy. The resting state is intermediate in energy and has a conformation that picks up a proton on one side of the membrane. It spontaneously converts to the second state at lower energy, shifting the conformation of the molecule and moving the proton to the other side. Absorption of light converts the complex into a third form of highest energy, which forces the release of a proton. The complex then spontaneously falls into the original intermediate-energy form, ready to pick up another proton. The cycle occurs only in one direction, pumping protons in one direction across the membrane, because of the need for light energy for the transition from low-energy to high-energy conformations. The key step is a change in conformation of the small cofactor molecule retinal, which is induced by absorption of light. A similar change in shape is used in visual sensing. Retinal contains a string of conjugated double bonds that absorb a wide range of visible light. When light is absorbed, one double bond switches from the straight *trans* form to the bent *cis* form. Retinal is connected covalently to the surrounding protein, so this large change in shape is transmitted to the protein, changing its conformation and promoting release of the proton. So far as F_oF_1-ATPase is concerned, bacteriorhodopsin could utilize light to translocate protons across the membrane, forming a proton gradient. Subsequently, the F_oF_1-ATPase complex employs the proton gradient to pump the protons into the cell and synthesize ATP. The adsorption spectra of bacteriorhodopsin ranges from 400 to 700 nm, with a peak absorbance at approximately 580 nm (yellow-green light).

Biomimetic systems consisting of bacteriorhodopsin and F_oF_1-ATPase in liposomes were first used to demonstrate the light-driven production of ATP. This artificial system reproduces ATP biosynthesis *in vitro*. Rigaud et al. studied the process of incorporating both bacteriorhodopsin and F_oF_1-ATPase into liposomes in detail [45, 47]. They first prepared liposomes with diameters on the scale of 100 nm in the bulk from purified lipids by selective solvent removal, using rotary evaporation. These large liposomes mimic a small cell, with the interior partitioned from the bulk by a lipid bilayer. In these systems, bacteriorhodopsin pumps protons from the surrounding media into the liposomes. The gradient is then consumed as F_oF_1-ATPase permits protons back across the liposome membrane, thus synthesizing ATP external to the liposome (Figure 3.5) [48]. Optimal condi-

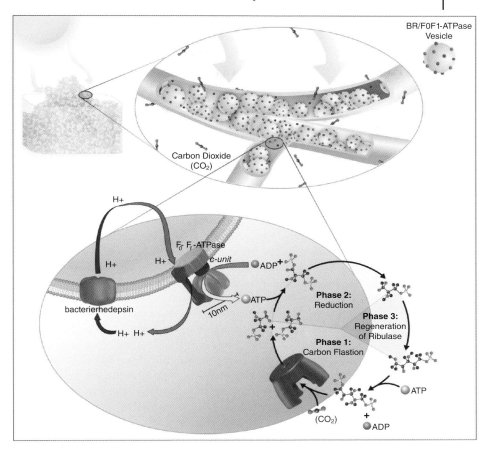

Figure 3.5 Illustration of a biomimetic solar conversion system formed by a vesicle containing both bacteriorhodopsin (BR) and F_oF_1-ATPase trapped within foam channels. Sunlight is converted into ATP which is then used by the Calvin–Benson–Bassham enzymes to make sugar from carbon dioxide and NADH (Reprinted with permission from [48]. © 2010, American Chemical Society).

tions for the reconstitution of both bacteriorhodopsin and F_oF_1-ATPase were obtained with octyl glucoside or Triton X-100 by insertion of the proteins into detergent-saturated liposomes. Subsequently, they found that final ATPase activities depended largely on the phospholipid/bacteriorhodopsin and the phospholipid/F_oF_1-ATPase ratios as well as the phospholipids used. Hence, a general and useful assay tool for the optimal reconstitution of bacteriorhodopsin with ATPase from different sources has been developed [49]. In the following sections, all proteoliposomes containing ATPase were prepared according to the optimal conditions described here. More recently, in order to address the instability issue of proteoliposomes, Montemagno *et al.* immobilized these proteoliposomes containing both bacteriorhodopsin and ATPase into a sol-gel matrix [50]. In their report, they incorporate these proteoliposomes into a colloidal suspension of silica particles with a porous structure (sol-gel matrix). The gel structure could prevent desiccation and also provides a more rigid stabilized system compared with a purely water aqueous environment. It can prevent protease access and keep vesicles trapped within the matrix. The studies show that proteins in the liposome/gel mixture system remained active for weeks, displaying a similar behavior to those incorporated in vesicle aqueous solution.

3.3.1.2 Proton Gradients Produced by Artificial Photosynthetic Reactions

The transmembrane electrochemical gradient plays a key role to generate energy "fuel" ATP molecules in living organisms. This gradient is created by redox reactions that are either photochemically driven, such as those in photosynthetic reaction centers, or intrinsically spontaneous, such as those of oxidative phosphorylation in mitochondria. The success of photosynthesis has prompted research on the natural process and spurred attempts to copy it in the laboratory. As we discussed in the previous section, bacteriorhodopsin can utilize sunlight to translocate protons across the membrane, forming a proton gradient. This is a prime example of solar energy conversion in biological systems. Increased understanding of photosynthetic energy conversion has made it possible to create artificial nanoscale devices and semibiological hybrids that carry out many of the functions of the natural process. However, bacteriorhodopsin is a biologically active protein and needs to be purified from other biological organisms. This means that bacteriorhodopsin is very expensive and unstable. In view of this, artificial photosynthetic centers have to be developed. In fact, until now, many approaches have been proposed to apply the basic physical and chemical principles of photosynthesis to artificial systems [51–55]. In this section, we mainly focus on the proton gradient produced by the artificial photosynthetic reaction, powering the generation of ATP molecules.

Moore *et al.* first reported an artificial photosynthetic membrane in which a photocyclic process was used to transport protons into a liposomal membrane, resulting in acidification of the liposome's internal volume [56–59]. This artificial photosynthetic reaction center comprises the carotenoid–porphyrin–quinone (C–P–Q) triad and quinone (Figure 3.6a). They were incorporated into liposomes containing F_oF_1-ATPase. These molecules are housed in the lipid bilayer membrane of a liposome vesicle with a diameter of around 150 nm (Figure 3.6b). The triad is inserted vectorially into the membrane so that the majority of the molecules have the quinone near the external surface. The proton transport is based on a redox loop. It is powered by the triad that, upon absorption of light by the porphyrin, generates a C^+–P–Q^- charge-separated state. The quinone radical anion of C^+–P–Q^- near the outer membrane surface reduces a molecule of quinone to yield the semiquinone anion, which is basic enough to accept a proton from the exterior aqueous environment. The resulting neutral semiquinone radical diffuses within the bilayer. When it encounters the carotenoid radical cation near the inner membrane surface, it is oxidized back to the quinone. The protonated quinone is

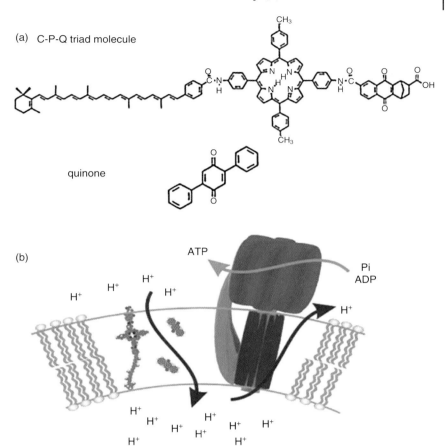

Figure 3.6 Molecular structure of C–P–Q triad molecules and quinone used as a synthetic photoreactive center. Schematic representation of an artificial photosynthetic membrane. The lipid bilayer of a liposome contains the components of a light-driven proton pump: a vectorially inserted C–P–Q triad molecule and "shuttle" quinone. Illumination of the triad leads to transport of hydrogen ions into the liposome interior, establishing a proton motive force. The membrane also contains a vectorially inserted ATPase enzyme. The flow of protons out of the liposome through this enzyme drives the production of ATP. (Reprinted with permission from [59]. © 1998, Nature Publishing Group.)

a strong acid that releases the proton into the inner volume of the liposome. It results in proton translocation into the vesicle interior and regeneration of the photoredox catalyst. Thus, the proton gradient across the membrane is created and the ATPase enzyme can then transport the protons back out of the liposome, synthesizing ATP molecules outside the liposome. Irradiation of this artificial membrane with visible light could mimic the process by which photosynthetic reactions in living organisms convert light into ATP chemical potential. The

3.3.2
ATP Biosynthesis from Biomimetic Microcapsules

3.3.2.1 Generation of Proton Gradients in Polymer Capsules by the Change of pH Values

Membrane-bound protein, F_oF_1-ATPase has been successfully reconstituted in liposomes acting as a biomimetic membrane, as we discussed in the Section 3.3.1. In these systems, bacteriorhodopsin utilizes light radiation to transfer protons from the surrounding media into the liposomes, generating a proton gradient. Subsequently, the gradient is consumed since F_oF_1-ATPase pumps protons back across the liposome membrane, synthesizing ATP outside liposomes. However, this is only a model system and liposomes with membrane proteins can hardly be considered a functional material due to the innate drawback of liposomes. In order to create useful materials, the system must be made more robust and be organized at scales larger than nanometers. Our group reported the reconstitution of F_oF_1-ATPase in the outer shell of lipid bilayer-coated polyelectrolyte microcapsules to imitate the motion in the living cell governed by molecular motors (Figure 3.7a) [60]. Microcapsules were first assembled by the alternating adsorption of negatively charged poly(acrylic acid) (PAA) (sodium salt) and positively charged poly(allylamine hydrochloride) (PAH) onto melamine formaldehyde particles as templates, followed by the removal of templates. The permeability of these layer-by-layer (LbL)-assembled PAA/PAH microcapsules can be readily tuned through the amount of assembled polyelectrolyte layers and other parameters (e.g., temperature, salt concentration, etc.) [61]. In particular, these assembled microcapsules can easily realize the biointerfacing by using the vesicle fusion method so that a new type of biomimetic structure – polyelectrolyte multilayer supported vesicles – can be formed [61–63]. This new system can greatly increase the vesicle stability and provide more possibilities to control the permeability of as-assembled capsules. Accordingly, the proton gradient can be created by the injection of buffer solutions with different pH values [64–66]. Changes in the pH value within the interior of the microcapsules can be assayed by using the fluorescence shift of the dye, pyranine (8-hydroxy-1,3,6-pyrenetrisulfonate) [67–69]. Pyranine can be encapsulated in the cavity of the capsules by mixing the dye and microcapsule solution (Figure 3.7b). The self-deposition of pyranines in the interior of capsules was derived from the charged species already existing within the interior of the capsules [61]. Internal pH in capsules can be measured by reading the characteristic peak intensities at 460 and 406 nm in the excitation spectrum since the relative fluorescence intensity ratio of pyranine at 406 and 460 nm is dependent on the proton concentration. The proton gradient is created by an acid/base transition between the interior and exterior of the capsules by the injection of different buffer solutions [70].

F_oF_1-ATPase-proteoliposomes were incubated with the polyelectrolyte microcapsule solution above the phase transition temperature of the phospholipids. The

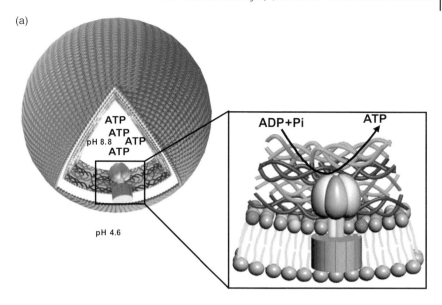

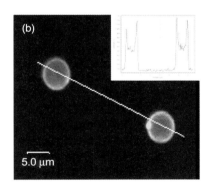
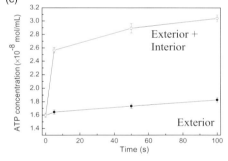

Figure 3.7 (a) Schematic representation of the arrangement of chloroplast F_oF_1-ATPase in lipid-coated microcapsules. (b) CLSM image of F_oF_1/lipid-coated $(PAA/PAH)_5$ capsules containing pyranine. Inset image: fluorescence intensity distribution along the line indicated in the confocal images. (c) ATP biosynthesis as a function of reaction time in chloroplast F_oF_1/lipid-modified polyelectrolyte microcapsules here, the "exterior" ATP content presents the produced ATP molecules in the solution except those embedded in the capsules. The "exterior+interior" means the total content of as-synthesized ATP molecules in both the solution and the capsules after the lipid bilayer was destroyed by addition of 0.1% triton X-100. (Reproduced with permission from Ref. [60]. © 2007 Wiley-VCH.)

proteolipids can fuse on the outer shell of the microcapsules due to the electrostatic interaction of phosphatidic acid with the cationic polyelectrolyte layer. Lipid-modified microcapsules with incorporation of F_oF_1 were thus obtained. To store as-synthesized ATP molecules inside the microcapsule, a proton gradient was generated by quickly injecting the same volume of acidic buffer solution.

The activity of F_oF_1-ATPase contained in the capsule outer shell can be detected by the measurement of ATP production by using the luciferin–luciferase system. This system is widely used for specific and sensitive estimation of ATP or rates of various processes synthesizing or consuming ATP [68–72]. This method is based on the fact that luciferase consumes ATP and produces light, which can be used to quantitatively determine ATP concentration. As the luminescence intensity is proportional to the amount of ATP present in the system, we first produced an ATP standard curve, with luminescence intensity versus ATP amount. Thus, the concentration of as-synthesized ATP in this biomimetic system can conveniently be obtained by reading the standard curve. ATP yield in capsules' exterior and interior solutions at different reaction times indicates that ATP content continuously increases and, more importantly, ATP concentration is clearly higher after the destruction of the lipid bilayers, indicating that ATP has been partly stored inside the microcapsules. The rate of ATP synthesis is also dependent on the magnitude of the proton gradient. In comparison with liposome-based ATP biosynthesis, we find out that LbL-assembled microcapsule-based ATP biosynthesis possesses several unique advantages in that the stepwise formation of microcapsules allows for easy manipulation of their properties, including size, shape, composition, and permeability (Figure 3.7c). The fact that F_oF_1-ATPase can be incorporated in lipid-modified microcapsules and retain their biological activity suggests that many other active membrane-bound proteins would be compatible with this kind of biomimetic microcapsule.

3.3.2.2 Proton Gradients in Protein Capsules Supplied by the Oxidative Hydrolysis of Glucoses

In the preceding section we described how we resolved the stability issue, reconstructing the ATP synthesis process in LbL-assembled polyelectrolyte microcapsules. The proton gradient can be generated by acid/base transition between the interior and exterior of the microcapsules. Coupling this proton gradient to ATPase reproduced the ATP-generating behavior of the real cell. However, the generation of the proton gradient cannot be maintained for a longer time through the injection of buffer solution with different pH values. A further challenge is how we can design a continuous proton gradient in the assembled biomimetic systems in such a way that we can make useful devices in the future. Our approach to this problem is utilizing the oxidative hydrolysis of the glucoses [73]. Energy from the transmembrane proton electrochemical gradient generated by oxidative metabolism is used by the mitochondrial ATPase to couple the proton motive force to the synthesis of ATP. It is well-known that the oxidative metabolism of glucose is used as an energy source in most organisms, from bacteria to humans. The oxidation of glucose is catalyzed by glucose oxidase (GOD) – a dimeric protein that catalyzes the oxidation of β-D-glucose into D-glucono-1,5-lactone with the consumption of molecular oxygen. The gluoconolactone can be further hydrolyzed to gluconic acid that releases protons. This process creates a proton gradient which drives the biological synthesis of ATP catalyzed by ATPase.

On the other hand, biocompatible hemoglobin (Hb) microcapsules produced by the LbL technique are of great interest because of their potential applications in medicine, catalysis, and cosmetics, as well as biotechnology. In most cases the walls from the reported microcapsules are composed of two or more polymer components, which are mainly bridged via electrostatic interaction. We have developed a method using glutaraldehyde (GA) as cross-linker via covalent bonding to fabricate (GA/Hb) microcapsules by the LbL technique [74–76]. GA has been proved to improve the permeability of Hb protein capsules in contrast to polyelectrolyte capsules. This is helpful to better tune the storage and release of encapsulated small molecules. The protein capsule provides a compartment, isolated from the outside world, limiting material transport and stabilizing membrane proteins. The next step was to reproduce the ATP-generating behavior of the liposomes. We have reported the reconstitution of F_oF_1-ATPase into lipid bilayer-coated Hb protein microcapsules and the ready production of the proton gradient can be readily produced by the hydrolysis of glucose catalyzed by GOD in the bulk solution of the outside capsules. Such a proton gradient can be maintained for a longer time with the addition of glucose. The generated proton gradients enable ATP synthesis in the biocompatible hollow protein capsules as illustrated in Figure 3.8. Similarly, to monitor internal pH changes induced by the catalytic oxidation of glucose, pyranine was entrapped inside the capsules by mixing the microcapsule suspension with pyranine. As in other experiments, proteoliposomes are formed with incorporated F_oF_1-ATPases. Then, F_oF_1-proteoliposomes were added and incubated for 30 min, followed by centrifuging and washing three times with buffer solution. Thus, pyranines were encapsulated inside the Hb protein capsules and ATPase-proteoliposomes were assembled onto the capsules shells.

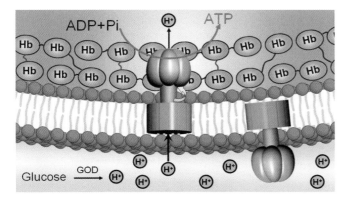

Figure 3.8 Schematic representation of ATP synthesis catalyzed by F_oF_1-ATPase reconstituted in lipid-coated Hb microcapsules.

Protons were released by adding glucose and GOD solution into ATPase/lipid-coated capsules suspension, forming a proton gradient between the interior and exterior of the microcapsules. As the relative fluorescence intensity ratio of pyranine at 406 and 460 nm is dependent on the hydrogen ion concentration, the fluorescence intensity at 460 nm decreases with time and at 406 nm increases, indicating that the solution inside the microcapsules was becoming increasingly acidic. This was the result of a continuous inward pumping of protons across the membrane. The inward pumping of protons provides a driving force for ATPase rotary catalysis, and ATP can thus be synthesized from $ADP + P_i$ in the solution. A standard luciferase assay was used to quantitatively determine the ATP amount. Results showed that the ATP concentration continuously increases with reaction time. More importantly, was obviously higher after the lipid membrane on the microcapsule was destroyed, indicating that ATP was mainly synthesized inside the microcapsules. Such assembled capsules are very stable under physiological conditions, especially in the presence of surfactants. Their longer lifetime will enhance the ATP production efficiency.

3.3.2.3 Proton Gradients Generated by GOD Capsules

Another strategy for generating a continuous proton gradient across the biomimetic microcapsules is to directly use GOD as block unit to fabricate microcapsules (Figure 3.9) [77]. Monitoring the pH change inside microcapsules indicates the continuous yield of protons in the GOD microcapsule suspension, confirming that GOD/GA microcapsules keep their catalytic activity. The quantitative exploration of enzymatic activity was performed on GOD cross-linked by GA on free GOD. It showed that a larger fraction of the GOD making up the wall is available for enzymatic reaction. Furthermore, the stability of GOD/GA microcapsules with respect to temperature and pH was also proved to be good. Confocal laser scanning microscopy (CLSM) images verified the successful adsorption of lipids onto the GOD/GA capsule shells. Moreover, internal pH changes demonstrate that protons are continuously produced in the interior of the GOD microcapsules. Once glucose solution was injected into the suspension, the hydrolysis of glucoses catalyzed by GOD creates a proton gradient between the outside and the inside the capsule walls, where chloroplast F_oF_1 ATPase is running and thus ATP synthesis from $ADP + P_i$ was started. The measurements reveal that ATP production continuously increases with time, corresponding to the continuous yield of protons. Interestingly, ATP can be partly synthesized in the interior solution. It may be attributed to part of the F_1 subunit extending into the capsules' interior aqueous phase. It is

Figure 3.9 (a) Biocatalytic oxidation process of β-D-glucose by GOD. (b) Schematic representation of the arrangement of chloroplast F_oF_1-ATPase in lipid-coated GOD capsule. (c) ATP biosynthesis as a function of reaction time in chloroplast F_oF_1/ lipid-modified polyelectrolyte microcapsules. Here, the interior solution represents the amount of released ATP after addition of 0.1 % Triton X-100. (Reprinted with permission from [77]. © 2009, American Chemical Society).

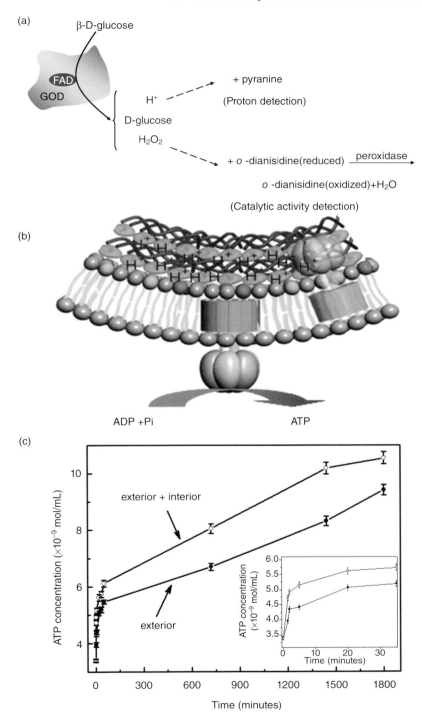

noted that in our work, no special steps are needed to orient the proteins. According to the described reconstitution procedure, most of the proteins were incorporated unidirectionally in small vesicles because they tend to preferentially insert with a particular orientation due to the shape of the protein molecule and its packing in the membrane. As the F_oF_1-proteoliposomes spread on the surface of the biomimetic microcapsules, formation of supported lipid bilayers on the microcapsules results in a more random orientation of the F_oF_1-ATPases.

As mentioned, no special steps are needed to orient the proteins in our work. According to the described reconstitution procedure, most of the proteins were incorporated unidirectionally in small vesicles because they tend to preferentially insert with a particular orientation due to the shape of the protein molecule and its packing in the membrane. As the F_oF_1-proteoliposomes spread on the surface of the biomimetic microcapsules, formation of supported lipid bilayers on the microcapsules results in a more random orientation of the F_oF_1-ATPases due to the difference of liposome fusion and the reverse of the protein (Figure 3.9). That is, the F_oF_1-ATPases were not embedded unidirectionally into the biomimetic microcapsule systems [78].

3.3.3
Reassembly of F_oF_1-ATPase in Polymersomes

As we discussed in the above section, phospholipid membranes are used in the preparation of spherical vesicles and planar membranes to study transmembrane protein activities. Reasonably, from an engineering viewpoint, one of the major properties required of membranes where proteins reside is chemical/mechanical stability that traditional lipid membranes do not have. Therefore, for the successful fabrication of membrane-bound hybrid devices, three fundamental parameters need to be satisfied: reassembly of artificial membrane/protein system, biocompatibility of the protein with membranes, and working environment.

Synthesized, amphiphilic block copolymers have been demonstrated to make more stable vesicles (polymersomes) than lipids (liposomes). The employment of synthetic polymers, in particular lipid-bilayer-like amphiphilic block copolymers, is intriguing because they are chemically and mechanically stable while still providing an amphiphilic structure that allows incorporation of membrane proteins [79–81]. The pioneering work of polymer membranes as a cellular mimetic architecture was performed by Meier et al. [81–83]. Since then, some membrane proteins have been functionally inserted into block copolymer membranes.

Recently, block copolymers have shown the stability needed for long-term systems, while also exhibiting biocompatibility with membrane proteins. These block copolymers could self-assemble to form a polymersome structure under mild conditions (Figure 3.10) [84]. This gives engineers the freedom to construct architectures in bulk solution. These preliminary studies are of increased importance in laying the foundation for the potential reconstruction of cellular processes in artificial systems for use in the fabrication of nanoscale organic/inorganic hybrid devices.

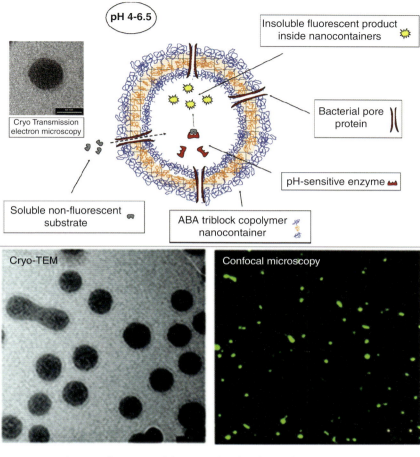

Figure 3.10 Schematic illustration of the nanoreactor system based on a self-assembled ABA triblock copolymer polymersome. This polymersome is functionalized with bacterial *ompF* pore proteins that make intact, size-selective channels for passive diffusion across the membrane. Encapsulated acid phosphatase enzyme could transform from a nonfluorescent, soluble substrate into an insoluble, fluorescent product at pH 4–6.5. (Reprinted with permission from [84]. © 2006, American Chemical Society.)

Based on the above research on protein-incorporated polymersomes, Montemagno *et al.* first applied a triblock copolymer (poly(2-ethyl-2-oxazoline)-*b*-poly(dimethylsiloxane)-*b*-poly(2-ethyl-2-oxazoline)) polymersome to incorporate ATPase generating ATP [85, 86]. This poly(dimethylsiloxane)-based ABA triblock copolymer was synthesized by ring-opening cationic polymerization of ethyl oxazoline with bifuctional benzyl chloride-terminated poly(dimethylsiloxane). This polymersome has a wall thickness of about 4 nm, which is similar to a typical lipid

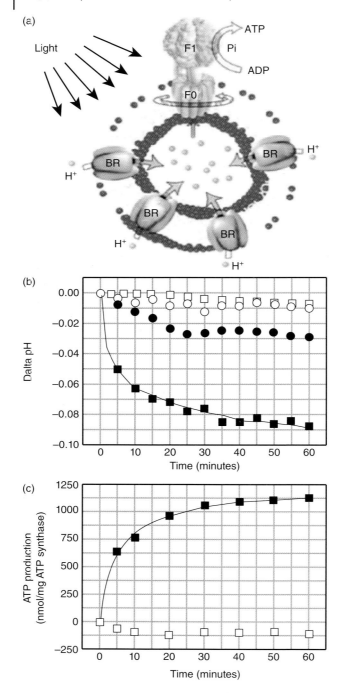

Figure 3.11 (a) Graphic representation of bacteriorhodopsin/ATPase polymersome. Polymersomes were prepared using poly(2-ethyl-2-oxazoline)-b-poly(dimethylsiloxane)-b-poly(2-ethyl-2-oxazoline) (PEtOz–PDMS–PetOz). (b) Proton-pumping activity of bacteriorhodopsin polymersomes monitored by trapping a fluorescent pH probe (pyranine) inside polymersomes. (c) Photoinduced ATP synthesis in bacteriorhodopsin/ATPase polymersomes. (Reprinted with permission from [85]. © 2005, American Chemical Society.)

bilayer thickness (around 5 nm). Obviously, it can serve as a replacement for traditional phospholipid bilayers. In order to create a proton gradient across the membrane, they still chose bacteriorhodopsin from *Halobacterium salinarium* as a light-driven proton pump. Bacteriorhodopsin and ATPase were then incorporated into the same polymersomes to generate ATP as schematically shown in Figure 3.11. The generation of a photoinduced electrochemical proton gradient from bacteriorhodopsin/ATPase polymersomes was measured by encapsulating a fluorescent pH probe dye inside the proteopolymersomes. The fluorescence pH measurements demonstrate that bacteriorhodopsin molecules retained their native photoactivity in the polymersomes. The production of ATP molecules in this system was quantitatively determined with a bioluminescence assay using the luciferin–luciferase method. The results show that ATP production markedly increased with the time of light irradiation. It also means that ATP can be generated from this bacteriorhodopsin/ATPase polymersomes through photoinduced phosphorylation. It proves the feasibility of synthesized polymersomes as a substitute for natural membranes in extremely demanding systems.

3.4
Conclusions and Perspectives

The biomolecular motor, F_oF_1-ATPase, can be assembled in different biomimetic systems and maintain high bioactivity. ATP production can be powered by a proton gradient across the microcapsule generated by a photoreactive reaction, acid/base transition, or glucose hydrolysis. In particular, these biomimetic vesicles or capsules can serve as containers for the storage of the synthesized ATP as a biological energy currency. By using this system, it becomes possible to study the function of ATPase in a biomimetic unit in detail. Furthermore, as vital activities need energy, ATP could be released from the assembled capsules to provide energy on demand.

References

1 Bao, G. and Suresh, S. (2003) Cell and molecular mechanics of biological materials. *Nat. Mater.*, **2**, 715–721.

2 He, Q., Cui, Y., and Li, J.B. (2009) Molecular assembly and application of biomimetic microcapsules. *Chem. Soc. Rev.*, **38**, 2292–2303.

3 Wendell, D.W., Patti, J., and Montemagno, C.D. (2006) Using biological inspiration to engineer functional nanostructured materials. *Small*, **2**, 1324–1329.

4 Lowe, C.R. (2000) Nanobiotechnology: the fabrication and applications of chemical and biological nanostructures. *Curr. Opin. Struct. Biol.*, **10**, 428–432.

5 Heuvel, M.G.L. and Dekker, C. (2007) Motor proteins at work for nanotechnology. *Science*, **317**, 333–338.

6 Junge, W. (1999) ATP synthase and other motor proteins. *Proc. Natl. Acad. Sci. USA*, **96**, 4735–4739.

7 Wang, H.Y. and Oster, G. (1998) Energy transduction in the F_1 motor of ATP synthase. *Nature*, **396**, 279–282.

8 Boyer, P.D. (1997) The ATP synthase: a splendid molecular machine. *Annu. Rev. Biochem.*, **66**, 717–749.

9 Sabbert, D. and Junge, W. (1997) Stepped versus continuous rotatory motors at the molecular scale. *Proc. Natl. Acad. Sci. USA*, **94**, 2312–2316.

10 Walker, J.E. and Dickson, V.K. (2006) The peripheral stalk of the mitochondrial ATP synthase. *Biochim. Biophys. Acta*, **1757**, 286–291.

11 Allison, W.S. (1998) F_1-ATPase: a molecular motor that hydrolyzes ATP with sequential opening and closing of catalytic sites coupled to rotation of its γ subunit. *Acc. Chem. Res.*, **31**, 819.

12 Boyer, P.D., Cross, R.L., and Momsen, W.A. (1973) New concept for energy coupling in oxidative phosphorylation based on a molecular explanation of the oxygen exchange reactions. *Proc. Natl. Acad. Sci. USA*, **70**, 2837–2839.

13 Sabbert, D., Engelbrecht, S., and Junge, W. (1996) Intersubunit rotation in active F-ATPase. *Nature*, **381**, 623–626.

14 Weber, J. and Senior, A.E. (1997) Catalytic mechanism of F_1-ATPase. *Biochim. Biophys. Acta*, **1319**, 19–58.

15 Duncan, T.M., Bulygan, V.V., Zhou, Y., Hutcheon, M.L., and Cross, R.L. (1995) Rotation of subunits during catalysis by *Escherichia coli* F_1-ATPase. *Proc. Natl. Acad. Sci. USA*, **92**, 10964–10968.

16 Boyer, P.D. (1989) A perspective of the binding change mechanism for ATP synthesis. *FASEB J.*, **3**, 2164–2178.

17 Boyer, P.D. (2000) Catalytic site forms and controls in ATP synthase catalysis. *Biochim. Biophys. Acta*, **1458**, 252–262.

18 Cox, G.B., Fimmel, A.L., Gibson, F., and Hatch, L. (1986) The mechanism of ATP synthase: a reassessment of the functions of the β and α subunits. *Biochim. Biophys. Acta*, **849**, 62–69.

19 Fimmel, A.L., Jans, D.A., Hatch, L., James, L.B., Gibson, F., and Cox, G.B. (1985) The F_1F_o-ATPase of *Escherichia coli*. The substitution of alanine by threonine at position 25 in the c-subunit affects function but not assembly. *Biochim. Biophys. Acta*, **808**, 252–258.

20 Howitt, S.M., Gibson, F., and Cox, G.B. (1988) The proton pore of the F_oF_1-ATPase of *Escherichia coli*: Ser-206 is not required for proton translocation. *Biochim. Biophys. Acta*, **936**, 74–80.

21 Jans, D.A., Fimmel, A.L., Langman, L., James, L.B., Downie, J.A., Senior, A.E., Ash, G.R., Gibson, F., and Cox, G.B. (1983) Mutations in the *uncE* gene affecting assembly of the c-subunit of the adenosine triphosphatase of *Escherichia coli*. *Biochem. J.*, **211**, 717–726.

22 Allison, W.S., Jault, J.-M., Grodsky, N.B., and Dou, C. (1995) A model for ATP hydrolysis by F_1-ATPases based on kinetic and structural considerations. *Biochem. Soc. Trans.*, **23**, 752–756.

23 Weber, J., Bowman, C., and Senior, A.E. (1996) Specific tryptophan substitution in catalytic sites of *Escherichia coli* F_1-ATPase allows differentiation between bound substrate ATP and product ADP in steady-state catalysis. *J. Biol. Chem.*, **271**, 18711–18718.

24 Menz, R.I., Walker, J.E., and Leslie, A.G. (2001) Structure of bovine mitochondrial F_1-ATPase with nucleotide bound to all three catalytic sites: implications for the mechanism of rotary catalysis. *Cell*, **106**, 331–341.

25 Gogol, E.P., Aggeler, R., Sagermann, M., and Capaldi, R.A. (1989) Cryoelectron microscopy of *Escherichia coli* F_1 adenosine triphosphatase decorated with

monoclonal antibodies to individual subunits of the complex. *Biochemistry*, **28**, 4717–4724.

26 Lücken, U., Gogol, E.P., and Capaldi, R.A. (1990) Structure of the ATP synthase complex (ECF_1F_o) of *Escherichia coli* from cryoelectron microscopy. *Biochemistry*, **29**, 5339–5343.

27 Abrahams, J.P., Leslie, A.G.W., Lutter, R., and Walker, J.E. (1994) The structure of F_1-ATPase from bovine heart mitochondria determined at 2.8 Å resolution. *Nature*, **370**, 621–628.

28 Ogilvie, I., Aggeler, R., and Capaldi, R.A. (1997) Cross-linking of the δ subunit to one of the three α subunits has no effect on functioning, as expected if δ is part of the stator that links the F_1 and F_o parts of the *Escherichia coli* ATP synthase. *J. Biol. Chem.*, **272**, 16652–16656.

29 Junge, W., Lill, H., and Engelbrecht, S. (1997) ATP synthase: an electrochemical transducer with rotatory mechanics. *Trends Biochem. Sci.*, **22**, 420–423.

30 Noji, H., Yasuda, R., Yoshida, M., and Kinosita, K. (1997) Direct observation of rotation of F_1-ATPase. *Nature*, **386**, 299–302.

31 Yasuda, R., Noji, H., Yoshida, M., Kinosita, K., and Itoh, H. (2001) Resolution of distinct rotational substeps by submillisecond kinetic analysis of F_1-ATPase. *Nature*, **410**, 898–904.

32 Mitome, N., Suzuki, T., Hayashi, S., and Yoshida, M. (2004) Thermophilic ATP synthase has a decamer *c*-ring: indication of noninteger 10:3 H^+/ATP ratio and permissive elastic coupling. *Proc. Natl. Acad. Sci. USA*, **101**, 12159–12163.

33 Adachi, K., Yasuda, R., Noji, H., Itoh, H., Harada, Y., Yoshida, M., and Kinosita, K. (2000) Stepping rotation of F_1-ATPase visualized through angle-resolved single-fluorophore imaging. *Proc. Natl. Acad. Sci. USA*, **97**, 7243–7247.

34 Nishizaka, T., Oiwa, K., Noji, H., Kimura, S., Muneyuki, E., Yoshida, M., and Kinosita, K. (2004) Chemomechanical coupling in F_1-ATPase revealed by simultaneous observation of nucleotide kinetics and rotation. *Nat. Struct. Mol. Biol.*, **11**, 142–148.

35 Soong, R.K., Bachand, G.D., Neves, H.P., Olkhovets, A.G., Craighead, H.G., and Montemagno, C.D. (2000) Powering an inorganic nanodevice with a biomolecular motor. *Science*, **290**, 1555–1559.

36 Bachand, G.D., Soong, R.K., Neves, H.P., Olkhovets, A., Craighead, H.G., and Montemagno, C.D. (2001) Precision attachment of individual F_1-ATPase biomolecular motors on nanofabricated substrates. *Nano Lett.*, **1**, 42–44.

37 Schmidt, J.J., Jiang, X.Q., and Montemagno, C.D. (2002) Force tolerances of hybrid nanodevices. *Nano Lett.*, **2**, 1229–1233.

38 Kaim, G. and Dimroth, P. (1998) Voltage-generated torque drives the motor of the ATP synthase. *EMBO J.*, **17**, 5887–5895.

39 Hisabori, T., Hara, S., Fujii, T., Yamazaki, D., Hosoya-Matsuda, N., and Motohashi, K. (2005) Thioredoxin affinity chromatography: a useful method for further understanding the thioredoxin network. *J. Exp. Bot.*, **56**, 1463–1468.

40 Itoh, H., Takahashi, A., Adachi, K., Noji, H., Yasuda, R., Yoshida, M., and Kinosita, K. (2004) Mechanically driven ATP synthesis by F_1-ATPase. *Nature*, **427**, 465–468.

41 Racker, E. (1979) Reconstitution of membrane processes. *Methods Enzymol.*, **55**, 699–711.

42 Sackman, E. (1996) Supported membranes: scientific and practical applications. *Science*, **271**, 43–48.

43 Stoeckenius, W. and Racker, E. (1974) Reconstitution of purple membrane vesicles catalyzing light-driven proton uptake and adenosine triphosphate formation. *J. Biol. Chem.*, **249**, 662–663.

44 Kagawa, Y., and Racker, E. (1971) Partial resolution of the enzymes catalyzing oxidative phosphorylation XXV. Reconstitution of vesicles catalyzing $^{32}P_i$-adenosine triphosphate exchange. *J. Biol. Chem.*, **246**, 5477–5487.

45 Rigaud, J., Pitard, B., and Levy, D. (1995) Reconstitution of membrane proteins into liposomes: application to energy-transducing membrane proteins. *Biochim. Biophys. Acta*, **1231**, 223–228.

46 Hokin, L.E. (1981) Reconstitution of "carriers" in artificial membranes. *J. Membr. Biol.*, **60**, 77–93.

47 Richard, P., Pitard, B., and Rigaud, J.-L. (1995) ATP Synthesis by the F_oF_1-ATPase from the thermophilic bacillus PS3 co-reconstituted with bacteriorhodopsin into liposomes. *J. Biol. Chem.*, **270**, 21571–21579.

48 Wendell, D., Todd, J., and Montemagno, C. (2010) Artificial photosynthesis in ranaspumin-2 based foam. *Nano Lett.*, doi: 10.1021/nl100550k.

49 Johnson, E.T. and Schmidt-Dannert, C. (2008) Light-energy conversion in engineered microorganisms. *Trends Biotechnol.*, **26**, 682–689.

50 Luo, T.M., Soong, R., Lan, E., Dunn, B., and Montemagno, C. (2005) Photo-induced proton gradients and ATP biosynthesis produced by vesicles encapsulated in a silica matrix. *Nat. Mater.*, **4**, 220–224.

51 Wasielewski, M.R. (1992) Photoinduced electron transfer in supramolecular systems for artificial photosynthesis. *Chem. Rev.*, **92**, 435–461.

52 Wagner, R.W. and Lindsey, J.S. (1994) A molecular photonic wire. *J. Am. Chem. Soc.*, **116**, 9759–9760.

53 Wagner, R.W., Lindsey, J.S., Seth, J., Palaniappan, V., and Bocian, D.F. (1996) Molecular optoelectronic gates. *J. Am. Chem. Soc.*, **118**, 3996–3997.

54 Kuciauskas, D., Liddell, P.A., Lin, S., Johnson, T.E., Weghorn, S.J., Lindsey, J.S., Moore, A.L., Moore, T.A., and Gust, D. (1999) An artificial photosynthetic antenna–reaction center complex. *J. Am. Chem. Soc.*, **121**, 8604–8614.

55 Meyer, T.J. (1989) Chemical approaches to artificial photosynthesis. *Acc. Chem. Res.*, **22**, 163–170.

56 Gust, D., Moore, T.A., and Moore, A.L. (2001) Mimicking photosynthetic solar energy transduction. *Acc. Chem. Res.*, **34**, 40–47.

57 Gust, D., Moore, T.A., and Moore, A.L. (1993) Molecular mimicry of photosynthetic energy and electron transfer. *Acc. Chem. Res.*, **26**, 198–208.

58 Steinberg-Yfrach, G., Liddell, P.A., Hung, S.C., Moore, A.L., Gust, D., and Moore, T.A. (1997) Artificial photosynthetic reaction centers in liposomes: photochemical generation of transmembrane proton potential. *Nature*, **385**, 239–241.

59 Steinberg-Yfrach, G., Rigaud, J., Durantini, E.N., Moore, A.L., Gust, D., and Moore, T.A. (1998) Light-driven production of ATP catalysed by F_oF_1-ATP synthase in an artificial photosynthetic membrane. *Nature*, **392**, 479–483.

60 Duan, L., He, Q., Wang, K.W., Yan, X.H., Cui, Y., Möhwald, H., and Li, J.B. (2007) ATP biosynthesis catalyzed by the F_oF_1-ATP synthase assembled in polymer microcapsules. *Angew. Chem. Int. Ed.*, **46**, 6996–6999.

61 Donath, E., Sukhorukov, G.B., Caruso, F., Davis, S.A., and Möhwald, H. (1998) Novel hollow polymer shells by colloid-templated assembly of polyelectrolytes. *Angew. Chem. Int. Ed.*, **37**, 2201–2204.

62 Decher, G. and Schlenoff, J.B. (2003) *Multilayer Thin Films*, Wiley-VCH Verlag GmbH, Weinheim.

63 He, Q., Möhwald, H., and Li, J.B. (2009) Layer-by-layer assembly of bioinspired composite nanotubes. *Curr. Opin. Colloid Interface Sci.*, **14**, 115–125.

64 Li, J.B., Möhwald, H., An, Z.H., and Lu, G. (2005) Molecular assembly of biomimetic microcapsules. *Soft Matter*, **1**, 259–264.

65 Ge, L.Q., Möhwald, H., and Li, J.B. (2003) Polymer stabilized phospholipid layers formed on polyelectrolyte multilayer capsules. *Biochem. Biophys. Res. Commun.* **303**, 653–656.

66 Ge, L.Q., Moehwald, H., and Li, J.B. (2003) Phospholipase A_2 hydrolysis of mixed phospholipid vesicles formed on polyelectrolyte hollow capsules. *Chem. Eur. J.*, **9**, 2589–2594.

67 Schmidt, G. and Graeber, P. (1985) The rate of ATP-synthesis by reconstituted CF_oF_1 liposomes. *Biochim. Biophys. Acta*, **808**, 46–54.

68 Wach, A., Dencher, N., and Gräber, P. (1993) Co-reconstitution of plasma membrane H^+-ATPase from yeast and bacteriorhodopsin into liposomes. ATP hydrolysis as a function of external and internal pH. *Eur. J. Biochem.*, **214**, 563–568.

69 Turina, P., Smaoray, D., and Gräber, P. (2003) H$^+$/ATP ratio of proton transport-coupled ATP synthesis and hydrolysis catalysed by CF$_o$F$_1$-liposome. *EMBO J.*, **22**, 418–423.

70 Possmayer, F. and Gräber, P. (1994) The pH$_{in}$ and pH$_{out}$ dependence of the rate of ATP synthesis catalyzed by the chloroplast H$^+$-ATPase, CF$_o$F$_1$ in proteoliposomes. *J. Biol. Chem.*, **269**, 1896–1904.

71 Capaldi, R.A. and Aggeler, R. (2002) Mechanism of the F$_1$F$_o$-type ATP synthase, a biological rotary motor. *Trends Biochem. Sci.*, **27**, 154–161.

72 He, Q., Duan, L., Qi, W., Wang, K.W., Cui, Y., Yan, X.H., and Li, J.B. (2008) Biosynthesis of ATP driven by the ATPase assembled in LbL-assembled microcapsules. *Adv. Mater.*, **20**, 2933–2937.

73 Qi, W., Duan, L., Wang, K.W., Yan, X.H., Cui, Y., He, Q., and Li, J.B. (2008) Motor protein CF$_o$F$_1$ reconstituted in lipid-coated hemoglobin microcapsules for ATP synthesis. *Adv. Mater.*, **20**, 601–604.

74 Qi, W., Yan, X.H., Fei, J.B., Wang, A.H., Cui, Y., and Li, J.B. (2009) Triggered release of insulin from glucose-sensitive enzyme multilayer shells. *Biomaterials*, **30**, 2799–2803.

75 Qi, W., Yan, X.H., Duan, L., Cui, Y., Yang, Y., and Li, J.B. (2009) Glucose-sensitive microcapsules from glutaraldehyde cross-linked hemoglobin and glucose oxidase. *Biomacromolecules*, **10**, 1212–1217.

76 Duan, L., He, Q., Yan, X.H., Cui, Y., Wang, K.W., and Li, J.B. (2007) Hemoglobin protein hollow shells fabricated through covalent layer-by-layer technique. *Biochem. Biophys. Res. Commun.*, **354**, 357–362.

77 Duan, L., Qi, W., Yan, X.H., He, Q., Cui, Y., Wang, K.W., Li, D.X., and Li, J.B. (2009) Proton gradients produced by glucose oxidase microcapsules containing motor F$_o$F$_1$-ATPase for continuous ATP biosynthesis. *J. Phys. Chem. B*, **113**, 395–399.

78 Naumann, R., Jonczyk, A., Kopp, R., van Esch, J., Ringsdorf, H., Knoll, W., and Gräber, P. (1995) Incorporation of membrane proteins in solid-supported lipid layers. *Angew. Chem. Int. Ed. Engl.*, **34**, 2056–2058.

79 Antonietti, M. and Forster, S. (2003) Vesicles and liposomes: a self-assembly principle beyond lipids. *Adv. Mater.*, **15**, 1323–1333.

80 Discher, D.E. and Eisenberg, A. (2002) Polymer vesicles. *Science*, **297**, 967–973.

81 Meier, W., Nardin, C., and Winterhalter, M. (2000) Reconstitution of channel proteins in (polymerized) ABA triblock copolymer membranes. *Angew. Chem. Int. Ed.*, **39**, 4599–4602.

82 Nardin, C., Hirt, T., Leukel, J., and Meier, W. (2000) Polymerized ABA triblock copolymer vesicles. *Langmuir*, **16**, 1035–1041.

83 Nardin, C., Thoeni, S., Widmer, J., Winterhalter, M., and Meier, W. (2000) Nanoreactors based on (polymerized) ABA-triblock copolymer vesicles. *Chem. Commun.*, **15**, 1433–1434.

84 Broz, P., Driamov, S., Ziegler, J., Ben-Haim, N., Marsch, S., Meier, W., and Hunziker, P. (2006) Toward intelligent nanosize bioreactors: a pH-switchable, channel-equipped, functional polymer nanocontainer. *Nano Lett.*, **6**, 2349–2354.

85 Choi, H. and Montemagno, C.D. (2005) Artificial organelle: ATP synthesis from cellular mimetic polymersomes. *Nano Lett.*, **5**, 2538–2543.

86 Choi, H.J., Lee, H., and Montemagno, C.D. (2005) Toward hybrid proteo-polymeric vesicles generating a photoinduced proton gradient for biofuel cells. *Nanotechnology*, **16**, 1589–1593.

4
Kinesin–Microtubule-Driven Active Biomimetic Systems

4.1
Introduction

Development of modern science and technology increasingly requires that alternative approaches to traditional engineering must be found to arrive at innovative solutions. One of the most promising of those approaches is molecular biomimetics. In fact, molecular biomimetics has proved very useful in the design and fabrication of new functional structured materials on the micro- and nanoscale. This technique is not limited to just learning inspiration from nature because, with the development of modern biology, scientists can directly utilize biological units themselves to construct hybrid nanostructured materials. Thus, some of the manufacturing difficulties of biomimetics can be avoided. Employing protein machines in a synthetic system is a good example of the latter.

Protein machines have been found to carry out tasks critical to cell function, including DNA replication, intracellular transport, ion pumping, and cell motility [1–4]. The linear molecular motors, such as kinesin, dyneins, and myosins, constitute a subset of these protein machines and are notable in being able to convert chemical energy directly to mechanical work [5–8]. In the cell, these motors generate the force that drives muscle contraction, transport intracellular cargo throughout cells, and determine the shape of each cell and, ultimately, the architecture of tissues and whole organisms. An impressive case of biological movement is the kinesin protein – a linear processive motor that transports chemical payloads along microtubules in the cell. The growing filaments used are microtubules that are polarized and have two different ends. These filaments also have a second and rather different function – they provide rails or tracks for the kinesin motors and the kinesin motors walk along the microtubules in a certain direction. As kinesin motors have evolved specifically to transport and organize material at the micro- and nanoscale, the integration of kinesin motors into engineered devices could overcome limitations in many fields of technology. Therefore, a major challenge for the construction of functional hybrid nanomaterials is how to integrate natural motor proteins into the engineering of biomimetic systems. However, a number of experimental problems such as interfaces between the proteins and material

Molecular Assembly of Biomimetic Systems. Junbai Li, Qiang He, and Xuehai Yan
© 2011 WILEY-VCH Verlag GmbH & Co. KGaA, Weinheim
ISBN: 978-3-527-32542-9

surfaces, and stability of the motor and microtubule proteins, must first been taken into account [9–12]. A number of insightful reviews have been written on the molecular mechanism of motor proteins and the application of biomolecular motors in nanotechnology [13–20]. In this chapter, we mainly focus on how molecular biomimetics applied to engineering kinesin–microtubule-based active biomimetic systems, particularly cargo selection, and loading and uploading of cargoes.

4.2
Kinesin–Microtubule Active Transport Systems

As mentioned above, we mainly focus on the kinesin–microtubule system because kinesin currently serves as a model protein for understanding the molecular basis of intracellular transport and for applications of molecular motors in nanotechnology. One of the principal functions of kinesin molecular motors is to transport vesicles along cytoskeletal filaments (i.e., microtubules) and to retain organelles at specific locations in the cell. Furthermore, they are also responsible for carrying out the bulk of intracellular transport. A great deal of structural information about kinesins is now available through crystal structures. Kinesins are a family of proteins that can be divided into 14 classes based on sequence similarity and functional properties. Almost all of the application work with kinesin motors has utilized conventional kinesin ("kinesin-1"). Other kinesins have different motor properties and they may become useful in future applications.

Kinesin-1 is a dimeric protein that contains three domains: the head or motor domain, the coiled-coil stalk that holds the two chains together, and the tail that, along with two associated light chains, is responsible for binding cargo. Each motor domain contains both an ATP- and a microtubule-binding site, and movement is achieved by a cycle in which each head alternately binds to the microtubule, undergoes a conformational change, and releases from the track. It has been demonstrated that when the kinesin walks, each of the motor heads is near the microtubule in the initial state, with each motor head carrying an ADP molecule to facilitate a stronger binding. Another ATP replaces the ADP, which leads to a conformational change. In the process, it pulls the other ADP-carrying motor head forward by about 16 nm so that it can bind to the next microtubule-binding site. The second head now binds to the microtubule by losing its ADP, which is promptly replaced by another ATP molecule. The first head hydrolyzes the ATP and loses the resulting P_i. It is then snapped forward by the second head while it carries its ADP forward. Hence, coordinated hydrolysis of ATP in the two motor heads is the key to kinesin processivity. Kinesin is able to take about 100 steps before detaching from the microtubule, while moving at roughly 800 nm s^{-1}.

The orientation of the "plus" and "minus" ends of microtubules establishes the direction of cargo movement. In most cells microtubules are nucleated at the centrosome, with their "minus" ends toward the center of the cell and "plus" ends at the cell periphery. Microtubules are cylindrical polymers of the protein tubulin that are 25 nm in diameter and up to tens of microns long. Tubulin dimers 8 nm long associate in a head-to-tail manner to make protofilaments; these protofilaments associate laterally to make sheets and the sheets close to make hollow cylinders that normally contain 13 protofilaments. As the subunits are asymmetric, microtubules exhibit an important structural property – the "minus" or slow-growing end is anchored near the center of the cell and the fast-growing "plus" ends extend to the perimeter of the cell. Experimentally, tubulin is normally isolated from cow or pig brains, which are large, inexpensive, and rich in tubulin due to the neurons that require long-distance intracellular transport. Microtubules can be polymerized *in vitro* from purified tubulin, covalently modified with fluorophores or other functional groups, and stabilized in polymer form with the drug taxol, making them stable.

4.3
Active Biomimetic Systems Based on the Kinesin–Microtubule Complex

Linear cytoskeletal kinesin motors have dominated the emerging field of protein-powered devices because they are relatively robust and readily available [21–27]. Tubulin can be commercially purchased, whereas the motor proteins can be purified from cells or expressed in recombinant bacterial systems and harvested in large quantities. Currently, kinesin–microtubule active biomimetic systems are fabricated mostly through two different approaches: microtubule gliding geometry and bead geometry.

4.3.1
Bead Geometry

Bead geometry is a commonly used assay because it is analogous to the geometry found in cells [28–32]. In this method, microtubules are immobilized on glass surfaces, motors are adsorbed to the beads, and the beads are transported along the microtubules (Figure 4.1). This requires the controlled placement of filaments onto a substrate and precoating of the cargo with motors. Often more motors are coated to a bead, which is helpful for long-distance transport and allows visualization of the movement or the exertion of force in an optical trap. In other words, multiple motors on a bead can attach to a single filament shuttle, so that large forces can be generated and sustainable movement can be realized. Several studies have demonstrated that optical tweezers can be used to grab the beads, and measure displacement and forces generated by the motors to nanometer and piconewton precision. The coupling of cargo to the motor protein is relatively straightforward. The precoating of kinesin motors onto the cargoes can be realized by using nonspecific electrostatic or hydrophobic adsorption or biomolecular recognition such as biotin–streptavidin or antibodies. For instance, the beads have been replaced by functional particles or biomolecules like proteins or nucleotides.

Figure 4.1 Schematic representation of bead geometry: microtubules are immobilized on glass surfaces, motors are adsorbed to the beads, and the beads are transported along the microtubules.

As the tail domain of kinesin motors can be deleted or significantly altered with no effect on the motor function, in theory antibody fragments, receptors, DNA-binding domains, or other protein motifs can be fused to the motor tail and these motor–cargo complexes transported along microtubules.

4.3.2
Gliding Geometry

In an alternative geometry, these motor systems are employed in a so-called microtubule gliding assay, in which the cytoskeletal filaments (usually about 1–20 μm in length) are propelled by surface-bound motors. Generally, kinesin motors are adsorbed to glass surfaces or microfluidic devices that have been treated with the blocking protein casein, and microtubules are observed landing on and moving over the motors (Figure 4.2). Using this approach, the motor concentration on the surface can be varied and different solutions can be introduced to optimize movement characteristics. The distance and trail also are independent on the microtubules. This is very important to design a complex device. The rotational flexibility of the motor stalks is high enough to rotate the randomly bound motors into the correct orientation for binding onto the microtubule. "Plus" end-directed motors will then propel the filaments with their "minus" end leading. Like nanoscale trucks, the microtubules can act as shuttles that transport an attached cargo such as nanoparticles or DNA.

Additionally, microtubule movements are visualized by covalently labeling the microtubules with a fluorescent dye, observing and recording them under a fluorescent microscope system. As the assay is relatively easy to perform and the filaments are transported long distances along the engineered surface, this approach has generated the most attention for the engineering of kinesin–microtubule-based biomimetic systems.

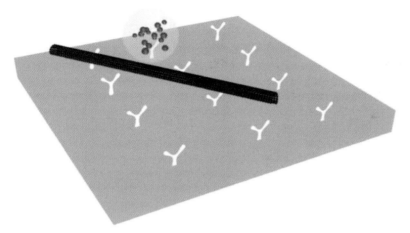

Figure 4.2 Schematic representation of microtubule gliding geometry: kinesin motors are immobilized to glass surfaces or microfluidic devices that have been treated with the blocking protein casein, and microtubules are observed landing on and moving over the motors.

4.3.3
Transport Direction and Distance of Assembled Systems

Motor directionality in either the gliding or bead geometry is a key consideration for many application fields such as sorting, separation, purification, assembly of materials, or drug delivery. These applications usually require the motion and transport of kinesin biomimetic systems independent of fluid flows or concentration gradients and can be unidirectionally guided along predesigned tracks. However, when motor proteins for the gliding geometry or microtubules for the bead geometry are adsorbed randomly onto the surface, the direction of cargo transport is random as well.

In the gliding geometry, microtubules diffuse out of solution and land on the motors, and the direction of microtubule transport can be defined by the orientation of the filament. As the motors, which are immobilized, move to the microtubule "plus" end, the filaments move with their "minus" ends leading. The coiled-coil of kinesin-1 has a region of random coil that enables the heads to rotate freely and bind to filaments only in the proper stereospecific orientation. Therefore, if motor proteins only adsorb onto the predesigned sites, the direction and distance of the microtubule movement can be guided so that unidirectional motion in the gliding geometry can be achieved. Hiratsuka *et al.* lithographically fabricated an arrowhead-shaped channel to selectively adsorb motor proteins [15]. The results demonstrated that this novel geometry of tracks can efficiently restrict microtubule movements along the tracks and further force microtubules to move in one direction. Populations of microtubules can move unidirectionally along linear tracks.

Other researchers extended this investigation using different fabrication techniques and a range of differently shaped channels. On the basis of this technique, microtubules are actively transported from one pool to the other in the direction of the arrowheads when two pools are connected by an arrowheaded track. Current study in this area aims to optimize the channel geometries to achieve ideal microtubule transport and redirect microtubule movement under an exerted field.

For the bead geometry, motors can easily be engineered to contain diverse cargo binding domains in place of their tail. Thus, binding motor proteins onto cargoes seems simpler. However, it has its own directionality problems – the transport direction is determined by the orientation of the immobilized microtubules. While filaments are aligned in fluid flows, their orientation is mixed. Hence, they are useless in transport applications. Recently, there has been progress in immobilizing aligned microtubules and transporting motor-functionalized cargo along these filaments. Hancock *et al.* prepared an array of uniformly oriented microtubules by fixing short microtubule seeds at defined locations, growing filaments selectively off of the "plus" ends of these seed, aligning the newly polymerized filaments by fluid flow, and then immobilizing them on a surface [21]. Prots *et al.* made a field of microtubules moving over immobilized kinesin motors, aligned their direction of movement using fluid flow, and then cross-linked the filaments to the motors [33]. Boehm *et al.* have shown that the flow of fluid can be used to orient microtubules into the same polarity and cargoes such as gold particles, glass, or polymer beads (1–10 μm) are unidirectionally translocated by kinesin motors [34]. Similarly, Hess, Vogel and others have developed several different kinesin-based nanodevices. One obviously disadvantage is that the transport distances are dependent on microtubule length and generally shorter than those obtained by the above-mentioned gliding method [14].

4.4
Layer-by-Layer Assembled Capsules as Cargo – A Promising Biomimetic System

4.4.1
Layer-by-Layer Assembled Hollow Microcapsules

It is well known that controlled encapsulation and release are extremely important in many fields such as food technology, medicine, drug delivery, coatings, and cosmetics [35–39]. In view of these applications, these capsule systems are required not only to have a cavity to encapsulate a variety of materials, but also to have capsules as well as capsule shells and interiors with different physiochemical properties. One of the most successful examples is the layer-by-Layer (LbL) assembly of polyelectrolyte nano- and microcapsules. The LbL technique was introduced at the beginning of the 1990s by Decher *et al.* (see Chapter 2). Originally this technique was based on the sequential adsorption of oppositely charged polymers (i.e., polyelectrolytes) on a charged planar substrate. Upon adsorption of a polyelectrolyte layer, charge overcompensation takes place, leading to a reversal of the surface charge, promoting the adsorption of the next, oppositely charged, polyelectrolyte. In this way one can easily prepare multilayered films with tunable physicochemical properties as both the number of layers as well as their composition can easily be varied. Currently, a large number of components, other than charged polymers, have been used to build multilayered films. DNA, proteins, nanoparticles, lipids, viruses, and so on, have been included in the multilayers, yielding thin films with tailor-made properties. Apart from electrostatic, other interactions such as hydrogen bonds, covalent bonds, biospecific interactions, stereocomplex formation, and so on, have also been used in order to accomplish an LbL build-up. Additionally, salt-, pH-, temperature-, glucose-, and biotin-responsive LbL films have been made by varying the chemical nature of the polyelectrolyte film. In 1998, Moehwald's group first utilized LbL assembly of oppositely charged macromolecules on removable colloidal particles to construct ultrathin hollow shells from nano- to microsize as schematically presented in chapter 2 Figure 2.1. These assembled capsules have well-controlled size, shape, and wall thickness. The wall composition can be readily changed to adjust their physicochemical properties and permeability. The capsule surface can be modified to alter functionality and/or improve the colloidal stability of the capsules, and various materials can then be sequestered into the capsule interior. Later, the capsules can release their content as a consequence of external or internal stimuli such as a change of pH, temperature, salt concentration, light irradiation, magnetic field, and biodegradability. These hollow capsules are often considered to have potential applications in the delivery and release of drugs, catalysis, biomedicine, and biomaterials.

Obviously, LbL-assembled capsules should be a promising cargo that can be conveniently integrated into the kinesin–microtubule systems.

4.4.2
Kinesin–Microtubule-Driven Microcapsule Systems

4.4.2.1 Fabrication in a Beaded Geometry

The microcapsule consisting of poly(styrene sulfonate) (PSS) and poly(allylamine hydrochloride) (PAH) was selected as a model carrier or cargo. Kinesin motors can be readily assembled on the surface of as-prepared capsules with PSS as the outmost layer via electrostatic interaction as well as likely involving nonspecific binding between PSS and kinesin. Rhodamine-labeled microtubules, composed of 13 protofilaments, were immobilized on the bottom of a flow chamber pretreated by an aminosilane solution through electrostatic attraction. While the kinesin-coated capsules were injected into the flow chamber, they had a chance to touch the immobilized microtubules. After addition of ATP into the flow chamber, as shown in Figure 4.3 the kinesin-coated capsules can walk from the "minus" to the "plus" end of the microtubules. ATP provides kinesin with energy to drive capsules moving along the microtubules. To drive the capsules, movement is correlated with the "hand-over-hand" stepping mechanism of kinesin motors. In this mechanism, one head of kinesin firmly binds to the track while the other head moves forward; the two heads alternate in taking steps from the "minus" to "plus"

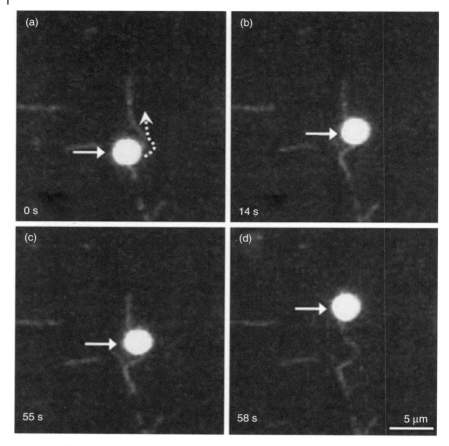

Figure 4.3 Time-lapse images of a capsule with coated kinesin motors moving along a microtubule. The solid arrows point out the moving capsules; the dotted white line indicates the direction of transport. (Reprinted with permission from [38]. © 2009, Elsevier.)

ends of microtubules [40]. This is why kinesin motors in most cells can drive cargos from the center of the cell towards the periphery. As such, the capsules with encapsulated dextran can also be transported by kinesin motors along microtubules, which is schematically shown in Figure 4.4. This indicates that the encapsulation of guests into capsules does not have a large influence on the transport of cargo along the microtubules. In some sense, this biomimetic system involving the driving of kinesin motors *in vitro* mimics intracellular vesicular transport.

4.4.2.2 Fabrication in a Gliding Geometry

As illustrated above, the assembled capsule can be a cargo to move along immobilized microtubules driven by kinesin motors. As an alternative avenue for the fabrication of such a class of active biomimetic systems, capsules with encapsu-

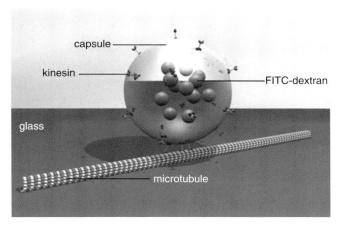

Figure 4.4 Capsule with encapsulation of guests as a cargo is driven by kinesin motors along a microtubule. (Reprinted with permission from [40]. © 2009, Elsevier.)

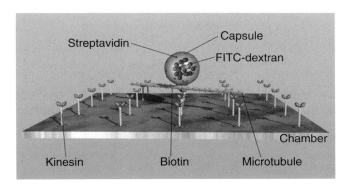

Figure 4.5 Transport of a polymer microcapsule filled with dextran by a microtubule on a kinesin-coated surface. The capsule is coated by streptavidins and thus bound to a biotinylated microtubule. Such a microtubule-carrying microcarrier can run on a kinesin-coated surface in the presence of ATP.

lated guests are designed to bind to the microtubules by using streptavidin–biotin specific recognition; then such a capsule–microtubule complex can be manipulated to run on a solid substrate coated with the kinesin motors (Figure 4.5). The merits of such an active device are that the capsules act as containers to carry various guest molecules or materials, and microtubules serve as the shuttles to possibly transport as-assembled capsules to the expected place or tissue for delivery application, biochemical sensing, and so on.

To achieve the integration of carriers with microtubules, the capsules are first coated with streptavidin and subsequently linked to the biotinylated microtubules.

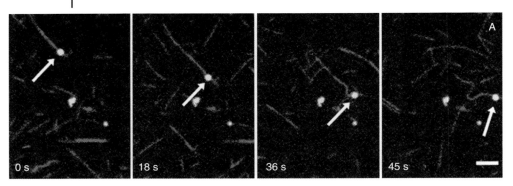

Figure 4.6 A microcapsule runs on the kinesin-coated surface. Scale bar = 5 μm. The moving microcapsule is indicated by white arrows.

Time-lapse images (Figure 4.6) show that with injection of ATP into the chamber some microtubules carrying capsules run on the kinesin-coated surface. The velocities of capsules are calculated by dividing run length by defined run time. The velocities of the microtubules carrying different capsules were 650–750 nm s^{-1} and thus not slower than the average cargo-free microtubule velocity [41]. It has been demonstrated that the average velocity of microtubules is independent of the diameters of attached capsules and the percentage of biotinylated tubulin. Intriguing, some steady capsules available in solution can be picked up by the passing microtubules and move together with them. A nearby microtubule without attached capsules moves in the same direction. The capsule is colliding with the passing microtubule and finally leaves for the solution. It thus is expected that these motility events have potential for the construction of microdevices.

4.5
Conclusions and Perspectives

The prepared polymer capsules are integrated with biological nanomachines in an artificial environment. In one case, capsules are able to move along the microtubules driven by kinesin motors similar to the intracellular transport of cargos. In addition, we likewise achieved the transport of capsules on the kinesin motor-coated surface via their complexation with microtubules. These kinesin motors provide the inspiration for the design and build-up of novel biomimetic functional nanomaterials. The integration of biomolecular motors with polymer microcarriers is important for handling multicomponent materials, which will pave the way to develop biomolecular motor-based hybrid microdevices with unique functionalities.

References

1 Schnitzer, M.J., Visscher, K., and Block, S.M. (2000) Force production by single kinesin motors. *Nat. Cell Biol.*, **2**, 718–723.

2 Hess, H., Clemmens, J., Brunner, C., Doot, R., Luna, S., Ernst, K.-H., and Vogel, V. (2005) Molecular self-assembly of "nanowires" and "nanospools" using active transport. *Nano Lett.*, **5**, 629–633.

3 Hess, H., Bachand, G.D., and Vogel, V. (2004) Powering nanodevices with biomolecular motors. *Chem. Eur. J.*, **10**, 2110–2116.

4 Caviston, J.P. and Holzbaur, E.L. (2006) Microtubule motors at the intersection of trafficking and transport. *Trends Cell Biol.*, **16**, 530–537.

5 Howard, J. (2001) *Mechanics of Motor Proteins and the Cytoskeleton*, Sinauer, Sunderland, MA.

6 Miki, H., Okada, Y., and Hirokawa, N. (2005) Analysis of the kinesin superfamily: insights into structure and function. *Trends Cell Biol.*, **15**, 467–476.

7 Vale, R.D. (2003) The molecular motor toolbox for intracellular transport. *Cell*, **112**, 467–480.

8 Vale, R.D., Reese, T.S., and Sheetz, M.P. (1985) Identification of a novel force-generating protein, kinesin, involved in microtubule-based motility. *Cell*, **42**, 39–50.

9 Seitz, A. and Surrey, T. (2006) Processive movement of single kinesins on crowded microtubules visualized using quantum dots. *EMBO J.*, **25**, 267–277.

10 Svoboda, K., Schmidt, C.F., Schnapp, B.J., and Block, S.M. (1993) Direct observation of kinesin stepping by optical trapping interferometry. *Nature*, **365**, 721–727.

11 Diehl, M.R., Zhang, K., Lee, H.J., and Tirrell, D.A. (2006) Engineering cooperativity in biomotor–protein assemblies. *Science*, **311**, 1468–1471.

12 Reuther, C., Hajdo, L., Tucker, R., Kasprzak, A.A., and Diez, S. (2006) Biotemplated nanopatterning of planar surfaces with molecular motors. *Nano Lett.*, **6**, 2177–2183.

13 Clemmens, J., Hess, H., Howard, J., and Vogel, V. (2003) Analysis of microtubule guidance in open microfabricated channels coated with the motor protein kinesin. *Langmuir*, **19**, 1738–1744.

14 Hess, H., Matzke, C.M., Doot, R.K., Clemmens, J., Bachand, G.D., Bunker, B.C., and Vogel, V. (2003) Molecular shuttles operating undercover: a new photolithographic approach for the fabrication of structured surfaces supporting directed motility. *Nano Lett.*, **3**, 1651–1655.

15 Hiratsuka, Y., Tada, T., Oiwa, K., Kanayama, T., and Uyeda, T.Q. (2001) Controlling the directions of kinesin-driven microtubule movements along microlithographic tracks. *Biophys. J.*, **81**, 1555–1561.

16 Moorjani, S.G., Jia, L., Kackson, T.N., and Hancock, W.O. (2003) Lithographically patterned channels spatially segregate kinesin motor activity and effectively guide microtubule movements. *Nano Lett.*, **3**, 633–637.

17 Hess, H., Clemmens, J., Howard, J., and Vogel, V. (2002) Surface imaging by self-propelled nanoscale probes. *Nano Lett.*, **2**, 113–116.

18 Xu, D., Watt, G.D., Harb, J.N., and Davis, R.C. (2005) Electrical conductivity of ferritin proteins by conductive AFM. *Nano Lett.*, **28**, 571–577.

19 van den Heuvel, M.G., de Graaff, M.P., and Dekker, C. (2006) Molecular sorting by electrical steering of microtubules in kinesin-coated channels. *Science*, **312**, 910–914.

20 Bachand, G.D., Rivera, S.B., Carroll-Portillo, A., Hess, H., and Bachand, M. (2006) Active capture and transport of virus particles using a biomolecular motor-driven, nanoscale antibody sandwich assay. *Small*, **2**, 381–385.

21 Muthukrishnan, G., Hutchins, B.M., Williams, M.E., and Hancock, W.O. (2006) Transport of semiconductor nanocrystals by kinesin molecular motors. *Small*, **2**, 626–630.

22 Hess, H., Clemmens, J., Qin, D., Howard, J., and Vogel, V. (2001) Light-controlled molecular shuttles made from motor proteins carrying cargo on engineered surfaces. *Nano Lett.*, **1**, 235–239.

23 Hess, H., Howard, J., and Vogel, V. (2002) A piconewton forcemeter assembled from microtubules and kinesins. *Nano Lett.*, **2**, 1113–1115.

24 Boal, A.K., Bachand, G.D., Rivera, S.B., and B.C. Bunker (2006) Interactions between cargo-carrying biomolecular shuttles. *Nanotechnology*, **17**, 349–354.

25 Diez, S., Reuther, C., Dinu, C., Seidel, R., Mertig, M., Pompe, W., and Howard, J. (2003) Stretching and transporting DNA molecules using motor proteins. *Nano Lett.*, **3**, 1251–1254.

26 Brunner, C., Wahnes, C., and Vogel, V. (2007) Cargo pick-up from engineered loading stations by kinesin driven molecular shuttles. *Lab Chip*, **7**, 1263–1271.

27 Ramachandran, S., Ernst, K.H., Bachand, G.D., Vogel, V., and Hess, H. (2006) Selective loading of kinesin-powered molecular shuttles with protein cargo and its application to biosensing. *Small*, **2**, 330–334.

28 Dinu, C.Z., Opitz, J., Pompe, W., Howard, J., Mertig, M., and Diez, S. (2006) Parallel manipulation of bifunctional DNA molecules on structured surfaces using kinesin-driven microtubules. *Small*, **2**, 1090–1098.

29 Du, Y.Z., Hiratsuka, Y., Taira, S., Eguchi, M., Uyeda, T.Q.P., Yumoto, N., and Kodaka, M. (2005) Motor protein nano-biomachine powered by self-supplying ATP. *Chem. Commun.*, 2080–2082.

30 Doot, R.K., Hess, H., and Vogel, V. (2007) Engineered networks of oriented microtubule filaments for directed cargo transport. *Soft Matter*, **3**, 349–356.

31 Beeg, J., Klumpp, S., Dimova, R., Gracia, R.S., Unger, E., and Lipowsky, R. (2008) Transport of beads by several kinesin motors. *Biophys. J.*, **94**, 532–541.

32 Liu, H.Q., Schmidt, J.J., Bachand, G.D., Rizk, S.S., Looger, L.L., Hellinga, H.W., and Montemagno, C.D. (2002) Control of a biomolecular motor-powered nanodevice with an engineered chemical switch. *Nat. Mater.*, **1**, 173–177.

33 Prots, I., Unger, E., Boehm, K.J. (2003) Isopolar microtubule arrays as a tool to determine motor protein directionality. *Cell Biol. Int.*, **27**, 251–253.

34 Boehm, K.J., Stracke R., Muhlig P., Unger E. (2001) Motor protein-dirven unidirectional transport of micrometer-sized cargoes across isopolar microtubule arrays. *Nanotechnology*, **12**, 238–244.

35 Konishi, K., Uyeda, T.Q., and Kubo, T. (2006) Genetic engineering of a Ca^{2+} dependent chemical switch into the linear biomotor kinesin. *FEBS Lett.*, **580**, 3589–3594.

36 Ionov, L., Stamm, M., and Diez, S. (2006) Reversible switching of microtubule motility using thermoresponsive polymer surfaces. *Nano Lett.*, **6**, 1982–1987.

37 Heuvel, M.G.L. and Dekker, C. (2007) Motor proteins at work for nanotechnology. *Science*, **317**, 333–336.

38 Browne, W.R. and Feringa, B.L. (2006) Making molecular machines work. *Nat. Nanotech.*, **1**, 25–35.

39 Hancock, W.O. (2006) In *Nanodevices for the Life Sciences* (ed. C. Kumar) Wiley-VCH Verlag GmbH, Weinheim, pp. 241–271.

40 Song, W.X., He, Q., Cui, Y., Möhwald, H., Diez, S., and Li, J.B. (2009) Assembled capsules transportation driven by motor proteins. *Biochem. Biophys. Res. Commun.*, **379**, 175–178.

41 Song, W.X., Möhwald, H., and Li, J.B. (2009) Triggered release of insulin from glucose-sensitive enzyme multilayer shells. *Biomaterials*, **30**, 2799–2806.

5
Biomimetic Interface

5.1
Introduction

Tools for nanofabrication have begun to provide important contributions for life sciences investigations, for developing biochip and biosensing technologies, as well as supplying basic research in lipid–protein, protein–protein, or polymer–protein interactions and protein function [1]. Scanning probe microscopy, scanning electron microscopy (SEM), and single-molecule fluorescence microscopy all supply tools for visualization and physical measurements [2]. Nanoscale studies can facilitate the development of new and better approaches for immobilization and bioconjugation chemistries, which are key technologies in manufacturing biochip and biosensing surfaces.

Biological or biomimetic interfacial patterning is essential for the integration of biological molecules into miniature bioelectronic and sensing devices [3–5]. To fabricate nanodevices for the life sciences it is often necessary to attach biomolecules to surfaces with retention of structure and function. For example, controlling the interaction of proteins, biomolecules, and cells with surfaces is important for the development of new biocompatible materials [6]. Precisely engineered surfaces can be used for the exploration of biochemical reactions in controlled environments. Spatially well-defined regions of surfaces can be constructed with reactive or adhesive terminal groups for the attachment of biomolecules [7]. Micropatterning of proteins has been applied for biosensors and biochips. Direct applications of protein patterning include biosensing, medical implants, control of cell adhesion and growth, and fundamental studies of cell biology [8–10]. Protein patterning has been accomplished at the micrometer level using microcontact printing [11, 12], photolithography [13], and microfluidic channels [14] (Table 5.1). In addition, microdisplacement lithography [15] or edge spreading lithography [16] with self-assembled monolayers (SAMs) allows lateral resolutions of about 100 nm. Recently, several types of lithographic processes have been used as research tools or in technical applications to structure surfaces at the nanometer scale (Table 5.1): optical lithography with UV or extreme UV radiation [20, 21], X-ray lithography [22],

Molecular Assembly of Biomimetic Systems. Junbai Li, Qiang He, and Xuehai Yan
© 2011 WILEY-VCH Verlag GmbH & Co. KGaA, Weinheim
ISBN: 978-3-527-32542-9

Table 5.1 Overview of the hierarchy of dimensions that can be achieved using various micro- and nanopatterning methods.

Method	Resolution	Area	References
Microcontact printing	~100 nm	>1 cm^2	[11–16]
Micro Displacement lithography			
Edge spreading lithography			
Nanoimprint lithography	~5 nm	>1 cm^2	[17]
Inkjet printing	~30 μm	>1 cm^2	[18]
Micellar lithography	~10 nm	>1 cm^2	[19]
Lithography with electrons, ions, and photons	~10 nm	>1 cm^2	[20–24]
EBL			
Electron projection lithography			
Focused ion beam			
Reactive ion etching			
Optical lithography (UV, extreme UV, X-ray)			
AFM/STM	~1 nm	nm^2–cm^2	[25–27]
Dip-pen nanolithography			
Nanografting			
STM lithography			

and lithography using particle beams [23] (e.g., electrons or ions). Nanoimprint lithography generates structures down to 5 nm [24]. The highest resolution can be achieved with serial techniques such as dip-pen nanolithography [25], nanografting [26] or scanning tunneling microscopy (STM) lithography [27]. However, these techniques are generally limited to small surface areas and are usually slow due to the serial writing mode. Several other special techniques such as micellar lithography [19], printing lithography (inkjet printing) [18], or stamp techniques such as nanoimprint lithography [17] and microcontact printing [28] are applicable to larger surfaces.

This chapter provides a brief overview of advances in the application of microcontact printing for lipid micropatterning, and electron beam lithography (EBL) for lipid nanopatterning and polymer gradient structures. In particular, in Section 5.2.3 and 5.3, a relatively new technique, chemical nanolithography, which is based on radiation-induced changes in organic SAMs, will be addressed.

5.2
Preparation and Characterization of Biomolecule Patterning

A number of factors need to be considered when choosing a successful protein-immobilization strategy, such as the efficiency and rate of binding, potential side-reactions, and the strength and flexibility of the attachment. Proteins have a three-dimensional structure that is critical to their function and activity. Most proteins have both positively and negatively charged regions that interact with surfaces. Upon encountering a surface, intramolecular forces within proteins can be disrupted, causing the proteins to unfold and become denatured [11]. Some proteins are known to lose activity when bound to a solid surface, due to a loss of tertiary structure. For retention of activity, chemistries for protein arrays should permit the immobilization of proteins on surfaces such that there is no perturbation to the native three-dimensional structure. The tools of organic chemistry provide a wealth of chemical strategies and binding motifs for conjugating biomolecules such as proteins to solid surfaces. Researchers have begun to use different strategies for linking proteins to surfaces through electrostatic interactions, covalent binding, or molecular recognition [2, 11]. The following sections introduce representative examples of immobilization strategies that have been applied for protein patterning.

5.2.1
Electrostatic Immobilization of Proteins for Surface Assays

The strategy of functionalizing a surface through electrostatic assembly is often used to immobilize biomolecules on surfaces. Electrically charged amino acids are found mostly on the exterior of proteins and can mediate assembly on charged surfaces [29, 30]. Proteins contain both positively and negatively charged domains that interact with surfaces via long-range electrostatic forces. The electrostatic attraction between oppositely charged molecules is nonspecific and surfaces are negatively or positively charged, depending on the solution pH. Electrostatic binding is physically mediated and proteins often retain their activity after immobilization. It is a direct, simple method for attaching proteins to surfaces without requiring multiple steps for chemical activation. Binding is reversible, since certain buffers and detergents can remove proteins from surfaces. However, a potential disadvantage of electrostatic immobilization is that the resulting orientation of proteins on surfaces is random. In other words, electrostatic-mediated binding does not provide a means for directing the protein assembly. This point can be partially overcome with the aid of several microfabrication techniques. For example, human serum albumin (HSA) has been used to generate a protein pattern on a negatively charged glass surface by using the microcontact printing technique [12, 31]. HSA has an isoelectric point at pH 4.8. Thus, HSA has a positive charge at pH 3.8. In view of this, a well-defined HSA micropattern has been fabricated by using microcontact printing. When HSA was labeled with 6-carboxyfluorescein (6-CF), we could clearly observe this structure.

5.2.1.1 Lipid-Modified HSA Patterns for the Targeted Recognition
From our previous studies [32–36], we have learned that the deposited film of HSA can interact with negatively charged lipid bilayers. Similarly, this routine was also carried out on a planar surface. Confocal laser scanning microscopy (CLSM) images confirmed the successful assembly of the HSA pattern and subsequent negatively charged lipid, L-α-dimyristoylphosphatidic acid (DMPA),

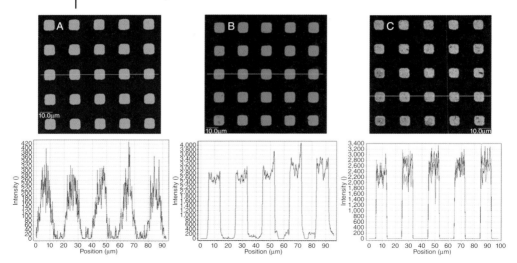

Figure 5.1 (a) Fluorescence image of the HSA patterns labeled by 6-CF constructed on a glass surface. (b) Fluorescence image of the lipid bilayers labeled by Texas Red–DHPE on HSA patterns. (c) Fluorescence image of the lipid bilayer (labeled by Texas Red–DHPE)/HSA (labeled by 6-CF) pattern system. Patterns are shown with scale bars and fluorescence profiles across the white lines drawn on images. (Reproduced with permission from [12]. © 2006, Elsevier.)

adsorption (Figure 5.1). By washing out the pattern plates, we did not find any change of fluorescence intensity, indicating that there is a stronger interaction between lipid and HSA. In this study, charge interaction and van der Waals force are the main interactions between protein and the glass. As for the adsorption of lipids, electrostatic force is dominant. In fact, the stability of these assembled protein or lipid/protein patterns is dependent on pH values. At pH 3.8, immersion time did not affect the intensity of fluorescence, whereas as the pH value increased to 7.4, the dipping weakened the intensity of lipid layers. In this case, the supported lipid membrane micropattern was fabricated. This is important because these constructs can serve as model systems for the description and investigation of membrane processes. Our group decorated lipid layers with single-stranded DNA (ssDNA) molecules so that the encoded targets could be recognized through DNA hybridization between complementary ssDNA chains. DMPA bilayers labeled by Texas Red–DHPE (1,2-dihexadecanoyl-*sn*-glycero-3-phosphoethanolamine, triethylammonium salt) and modified with ssDNA and without ssDNA, respectively, were immobilized in HSA patterns. When polystyrene microparticles coated with lipid layers containing the complementary ssDNA were incubated with the above-mentioned two patterns, respectively, these lipid-modified polystyrene microparticles deposited on the surface of ssDNA-modified DMPA/HSA patterns. In contrast, after the normal washing procedure no polystyrene microparticles were observed on the surface of DMPA/HSA patterns without ssDNA modification. This shows that the hybridization of ssDNA plays a key role.

5.2.1.2 Lipid-Modified HSA Patterns for *Escherichia coli* Recognition

It is well known that biomembrane surfaces contain a number of glycolipids and glycoproteins acting as intermediates for biological communications and recognition processes [37, 38]. Bacterial adhesion on the biomembrane surface will eventually cause infection. Therefore, the investigation of such adhesions accompanying recognition will help us to understand the mechanism of interaction to establish a rapid detection method for *E. coli*. Glycolipid arrays play a key role in a rapid detection and classification method for pathogens based on their adhesion profile. Several efforts have recently been made to study carbohydrate–protein interactions and to detect *E. coli* [39–43]. In view of performing molecular recognition based on *N*-acetylglucosamine groups, our group has recently combined the microcontact printing technique and liposome fusion to obtain organized glycolipid molecular patterns to recognize *E. coli* as illustrated in Figure 5.2 [12]. Similarly, HSA patterns were first prepared on a clean glass surface through the microcontact printing technique and then the lipid layer could be obtained by spreading DMPA vesicles containing glycolipid, 10-tetradecyloxymethy-3,6,9,12-tetraoxahexacosyl 2-acetamido-2-deoxy-β-D-glucopyranoside (PB1124), on HSA patterning [44]. Protein patterning and subsequent lipid adsorption were confirmed by using CLSM. When a printed microarray with glycolipid/DMPA was incubated with a solution of *E. coli*, the *E. coli* bacteria expressed by Green Fluorescent Protein can selectively adhere on the microarray via glycolipid (Figure 5.3). Reasonably, the mass fraction of the glycolipid inserted in the bilayer matrix pattern influences the absorption number of *E. coli* cells. The microarray becomes an *E. coli* bacteria pattern. The whole pattern surface is fully covered by the *E. coli* bacteria. These results further verified the specific interaction between *E. coli* and glycolipid. The *E. coli* recognition experiment performed under different time intervals shows that different amounts of bacteria are bound to the printed glycolipid surface. With the increase of incubation time, the number of *E. coli* on patterns is increased accordingly. In order to detect the binding of *E. coli* bacteria with glycolipid as a unique recognition, we carried out the experiment by treating the patterns at different pH values. In LB medium or in pH 7.0 phosphate buffer, we did not observe typical *E. coli* bacteria adhesion on the substrate. However, as we adjusted the pH value to 5.5 with a citric acid buffer under the same conditions, *E. coli* bacteria adhesion appeared. It is supposed that the conformation of the *E. coli* cell surface changes with pH values. To investigate further whether this *E. coli* strain could respond specifically to the interaction to the glycolipid, we performed another experiment using DMPA/HSA patterns without glycolipids. The results indicate that such patterns cannot give rise to the adsorption of *E. coli*. This indicates that the binding of *E. coli* to glycolipid is specific. Such patterns are expected to extend the technique of bacterial detection with a higher resolution.

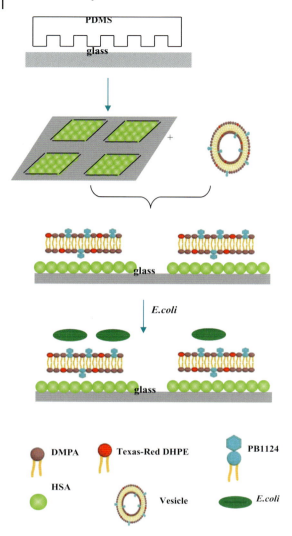

Figure 5.2 Schematic representation of *E. coli* cell adsorption on glycolipid/HSA patterns.

5.2.2
Covalent Immobilization of Proteins

Covalent immobilization is important for applications in which displacement or desorption of proteins can be a problem. Covalent bonds occur when two molecules share atoms and form the strongest chemical bonds for surface immobilization. The methods of covalent attachment are boundless – thousands of proteins have been immobilized on hundreds of different solid supports for affinity-capture

5.2 Preparation and Characterization of Biomolecule Patterning | 109

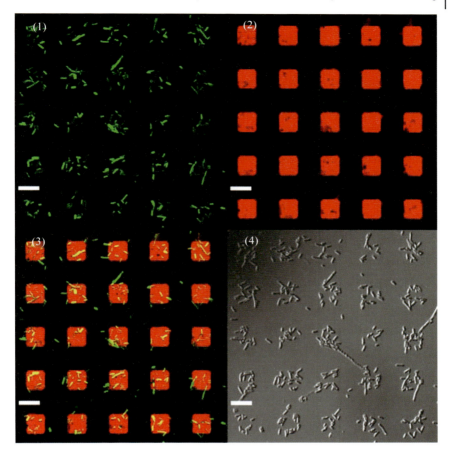

Figure 5.3 CLSM images of a glycolipid/HSA microarray after incubation with *E. coli* bacteria in 1:10 PB1124/DMPA ratios. (1) Fluorescence image of Green Fluorescent Protein-expressing *E. coli* bacteria confined to the lipid patterns; (2) fluorescence image of the lipid (labeled by Texas Red–DHPE) patterns supported by HSA; (3) overlap fluorescence image of (1) and (2); (4) light image of *E. coli* bacteria adsorbing on the lipid patterns. Scale bar = 10 μm. (Reproduced with permission from [31]. © 2007, Elsevier.)

assays [45–48]. The quality of covalent immobilization will depend on the chemical properties of both protein and the surface. Several amino acids provide suitable functional groups (i.e., amino groups, carboxyl groups, sulfhydryl groups, hydroxyl groups, and phenyl groups) for covalent modification. Since proteins typically present a number of these groups, the chemical nature of the support surface becomes a primary consideration. A specific chemical reaction is chosen to activate the surface and then proteins are immobilized upon exposure to the active surface groups. Examples of chemistries for covalent immobilization of proteins include activation of surface hydroxyl groups, carboxyl groups, and amines. Furthermore, bifunctional cross-linking reagents such as glutaraldehyde have been used to covalently couple proteins to various surfaces [30].

Importantly, covalent immobilization of proteins can be conveniently combined with other nano- or microfabrication techniques because controlled and accurate immobilization of a small number or even individual biomolecules at a specific location is also a key issue in biotechnology. In this case, chemically defined surface nanostructures are advantageous as they offer laterally well-defined sites at which biomolecules can be covalently attached to specific functional elements. Materials that are well suited to build surface nanopatterns are SAMs [49, 50]. SAMs are homogeneous, highly ordered films of organic molecules covalently anchored to a solid surface. In general, SAM systems can be tailored to bind to the surfaces of noble metals, semiconductors, and oxides. In biological applications, SAMs are frequently used as linkers to attach cells and proteins to surfaces. One of the most powerful techniques to generate nanoscopic laterally patterned SAMs is EBL. Recently, Gölzhäuser et al. have developed a simple routine to generate chemically distinct surface nanostructures based on SAMs combined with EBL [51–54]. Briefly, a densely packed monolayer of 4'-nitro-1,1'-biphenyl-4-thiol (NBT) was self-assembled on a gold surface. Electron beam irradiation was used to locally modify the terminal nitro groups to amino groups, while the aromatic layer was dehydrogenated and cross-linked. Subsequently, proteins such as streptavidin were covalently bound on the predefined sites. At the same time, nonspecific protein adsorption was prevented by binding of densely packed poly(ethylene glycol) (PEG) to the remainder of the surface [55]. They used glutaraldehyde as cross-linker to covalently attach proteins to the amino-modified regions of electron beam-patterned NBT SAMs (Figure 5.4).

5.2.3
Covalent Immobilization of Lipid Monolayers

Supported lipid membranes on solid substrates have been used to provide models mimicking a natural environment to study molecular recognition, cell adhesion, intercellular communication, and enzymatic catalysis [56–59]. Patterned lipid monolayers or bilayers are suitable artificial biomembrane models to integrate various components into the membrane system with defined spatial control and mimic the sophisticated functions of real biological membranes. Box et al. have reported micropatterning of supported lipid membranes by using different approaches including modification of the substrate, the microcontact printing technique, and so on. [60–62]. In relation to the study of membrane biophysics and biotechnology, supported lipid membranes at the nanoscale size have specific importance. However, the above-mentioned methods have some difficulties in obtaining nanoscale patterns. Recently, our group cooperated with Grunze's group and presented a method for fabricating lipid monolayer patterns through chemical bonding on SAMs by electron beam etching from micro- to nanometer size [63].

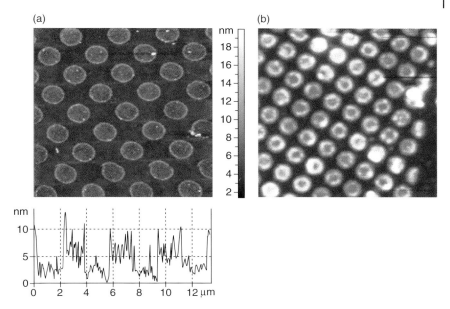

Figure 5.4 AFM (a) and fluorescence intensity image (b) (20 μm × 20 μm, 100 kW cm^{-2}, intensity scale: 0–180 counts/3 ms) of streptavidin-treated 1.5-μm patterns generated by EBL in NBT after cross-linking with a second layer of Cy5-biotin-labeled streptavidin. The height of the protein spots is around 4 nm and nonspecific adsorption of proteins on the surface is very low. The fluorescent spots show S/B ratios ~2. (Reproduced with permission from [55]. © 2004, Elsevier.)

SAMs have been used as resists in electron beam, UV, and X-ray lithography to produce chemical patterns, which serve as templates for proteins and other biosystems. As mentioned above, we utilized electron beam-induced decomposition and polymerization in NBT SAMs on gold to obtain micro- or nanopatterned amino structures on solid substrates [51–55]. Electron irradiation of NBT SAMs through a suitable mask or direct writing results in selective and quantitative reduction of the nitro groups to amino groups, while the underlying aromatic biphenyl layer is dehydrogenated and cross-linked. The amino moiety in the films can subsequently be used for selective binding of functional entities ("chemical nanolithography"). For high-resolution patterning, electron beams can be focused to nanometer-sized spots and the resolution is only limited by secondary electrons that are generated during irradiation. The smallest SAM structures that have been generated with electron beams are around 5 nm in size and the smallest chemical nanostructures obtained by chemical lithography had lateral dimensions of around 10 nm. The selectivity, flexibility, high throughout, and superior resolution of chemical lithography make it an ideal technological platform for the preparation of functionalized SAM patterns that can be amplified by chemical reactions. The basic strategy of our patterning and subsequent interfacial chemical reactions is depicted in Figure 5.5(A). Briefly, the terminal amino groups were firstly diazotized at around 0 °C by treatment with an aqueous solution of HCl (2.5 mol l^{-1} hydrochloric acid) and subsequent treatment with 0.5 mol l^{-1} sodium nitrite water solution. The samples with the diazonium sites were immersed in a 1 mM DMPA methanol solution to promote electrostatic assembly. In the self-assembly process, the diazonium (N^{2+}) and DMPA headgroups will bind via electrostatic attraction. Finally, the film was exposed to UV light (wavelength = 365 nm) for 30 min to ensure the photoreaction of diazonium with the phosphate group. The diazonium groups interacting with the phosphate groups of DMPA molecules via electrostatic attraction decomposed and released N$_2$ upon UV irradiation to produce phenyl cations, which then underwent an S$_N$1 type of nuclear displacement by the phosphate in DMPA headgroups. The photoreaction changes the interactions between DMPA and the diazonium group in the patterned SAM from ionic to covalent, and greatly improves the stability of phospholipid SAM. Before and after the reaction of DMPA with the diazonium functionalized gold surface, the presence of a P (2p) peak surface probed by X-ray photoelectron spectroscopy (XPS) provided direct evidence that the lipids are indeed covalently bonded at the surface, since any unbound moieties should have been removed by washing and sonication in dimethylformamide, ethanol, and chloroform. Atomic force microscopy (AFM) of DMPA monolayer patterning shows that the monolayer has an average thickness of 2.5 ± 0.1 nm (Figure 5.5B). Compared with the estimated lipid dimension (2.7 ± 0.1 nm in length, i.e., total bond lengths), the molecules in the monolayer appear to be tilted by an angle of 25–30 °C to the surface normal as observed in alkyl chain SAM systems according to X-ray diffraction results [64]. The result is close to the expected DMPA monolayer thickness measured in Langmuir–Blodgett films [65]. The immobilized DMPA was limited to the NBT SAM originally written by EBL, resulting in a one-to-one translation of the mask pattern in the nanometer range. The resulting lipid-modified molecular patterns are considered to be a first step towards obtaining stable biointerfacing patterns and studying biomolecular recognition.

Figure 5.5 (A) Schematic diagram of electron-induced chemical lithography and subsequently chemical reaction. First, an electron beam converts the terminal nitro groups of a NBT monolayer to amino groups while the underlying aromatic layer is cross-linked. Second, diazotization and electrostatic assembly with DMPA solution gives a patterned DMPA SAM. Finally, by exposure to UV light for a given time, the photoreaction results in a covalent bond between DMPA and the patterned substrate. (B) AFM micrographs of DMPA patterns that were generated by irradiation of a NBT SAM with low-energy electrons and subsequent chemical reactions. The height profiles show an average line-width around 250 (a) and 100 nm (b), and the average height of each line is about 2.5 nm. (Reproduced with permission from [63]. © 2004, American Scientific Publishers.)

5.2 Preparation and Characterization of Biomolecule Patterning

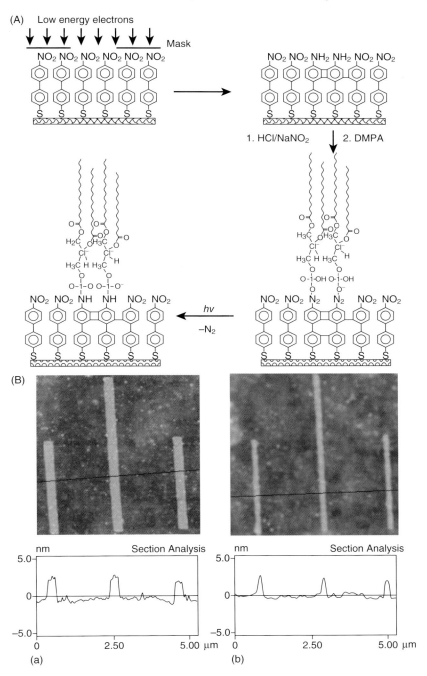

5.3
Polymer Brush Patterns for Biomedical Application

Over the last three decades, the importance of biomaterials has been recognized for biomedical applications and development of biomaterials has ranged from materials synthesized from metals, ceramics, or polymers to biological materials [66]. Generally, commercial-grade biomaterials may have excellent bulk properties, such as strength and elasticity, but they often show relatively poor surface properties, such as poor wear resistance and limited biocompatibility. The critical factors for biocompatibility are wettability (hydrophilicity/ hydrophobicity), chemistry, charge, and roughness. Consequently, modification of biomaterials by surface treatment has been an important focus for decades in surface engineering [7, 67, 68]. In particular, novel methods for the synthesis and fabrication of nanopatterned polymer or biomolecular brushes, with controlled molecular architecture, chemical functionality, and size, have potential for tailoring surface properties by imparting them with desirable energetic, mechanical, biological, optical, and electrical properties [69–73]. The lateral conformational restraint of polymer brushes on a surface leads to steric repulsive interactions between the packed chains and causes the polymer molecules to extend away from the substrate surface. Such brushes are often confined to predefined sites on nanostructured substrates. Also, polymers and biomacromolecules can undergo phase transitions with measurable structural and morphological changes upon variations to external stimuli such as temperature, ionic strength, and pH value [74–76]. Obviously, these environment-responsive polymer brush nanopatterns are important within a wide range of applications, such as biosensors, proteomic chips, and nanofluidic devices. Generally, two synthetic approaches have been used to synthesize polymer brushes at a nanostructured substrate: the "grafting-from" approach, through surface initiated polymerization (SIP), which yields polymer brush structures by direct synthesis from the surface, and the "grafting-to" approach that binds presynthesized polymers or biomacromolecules to a surface [77, 78]. A main drawback of the grafting-to method lies in the limited grafting density that can be achieved due to the increasing steric hindrance as the brushes attach to the surface. In SIP, however, the polymer chains are grown from surface-bound initiators, eliminating much of the steric effects that are associated with attaching polymer chains to the surface. When compared with the grafting-to approach, this grafting-from approach provides greater control over grafting density and yields higher packing densities. SAMs, such as suitably functionalized alkylchlorosilane and alkanethiol SAMs, provide reliable initiators for SIP [77].

5.3.1
Thermosensitive Polymer Patterns for Cell Adhesion

Currently, polymer brush nanostructures have been fabricated from the bottom up by using a variety of SIP methods, including anionic, cationic, plasma-induced, condensation, photochemical, electrochemical, and ring-opening metathesis polymerization [76]. However, the preparation of precisely patterned, surface-bound polymeric nanostructures with controlled lengths, conformational geometries, functionalities, and properties is still a challenging issue. In particular, atom transfer radical polymerization (ATRP) affords advantages over other synthetic methods [79]. It produces well-controlled brush growth under mild conditions, with defined molecular weight and low polydispersity. Importantly, the growth rate of polymer brushes can be well tuned by altering the ratio and concentration of reactants and catalysts. ATRP in water media at room temperature will be beneficial to prepare grafted polymer brushes on thiol SAM. Recently, our group in cooperation with Grunze's group has utilized the combination of chemical lithography and consecutive surface-initiated (SI)-ATRP to prepare various polymer brush nanostructures because this combination allows superior control of pattern formation and amplification of the patterns with a wide choice of monomers and chemical functionalities [80].

Poly(N-isopropylacrylamide) (PNIPAM) has been chosen as a model polymer because it is a well-known thermal-sensitive polymer that possesses a lower critical solution temperature (LCST) of around 32 °C in pure water [81]. Polymer brush patterns are fabricated on the basis of the procedures described in Figure 5.6(A). The reflected Fourier transform IR (FTIR) spectra and XPS data have been used to confirm the formation of the surface initiator monolayer as well as the existence of the PNIPAM brush. SEM and AFM were also used to observe the successful preparation of the nanopatterned polymer brushes (Figure 5.6B). Interestingly, PNIPAM brush patterns with different line widths from micro- to nanometer size show that compared with the grafting region, the measured widths of polymer brushes have an obviously expansive behavior, but the polymerization still exhibits linear growth. Although the polymerization conditions and the electron doses used were identical, the brush height was higher than those of the PNIPAM brushes of micrometer size shown above obtained with the flood gun at lower energies. The differences in brush height may be due to an energy-dependent cross-section for reduction of the nitro groups or to the higher yield of secondary electrons. To test the external stimuli of the PNIPAM patterns, thermosensitivity of the patterns was measured by observing the change of contact angle of PNIPAM film on a flat substrate with temperature. The AFM images in a water phase show that the height of the PNIPAM pattern at room temperature (below the LCST) is relatively high. Above the LCST, the PNIPAM height was greatly reduced, indicating that water in the PNIPAM film was expelled and that the polymer collapsed, which leads to the more hydrophobic PNIPAM-grafted surface (Figure 5.7).

In addition, many research groups have extensively studied the effects of surface wettability on the interactions of biological species with surfaces [82, 83]. The interactions of biological species with polymer surfaces mainly concern phenomena such as protein adsorption and cellular adhesion and spreading that take place in a biological environment. The adhesion and proliferation of different types of mammalian cells is also influenced by polymer surface wettability. Thus, the above-mentioned environmentally responsive PNIPAM brush can mimic the extracellular matrix and further help study the behavior of cell adhesion. Cell

116 | *5 Biomimetic Interface*

(A)

Figure 5.6 (A) Representation of the preparation process of PNIPAM brushes by chemical lithography and SI-ATRP on gold. (B) (a) and (b) SEM micrographs of PNIPAM brushes generated by chemical lithography and subsequent SI-ATRP on a substrate that was irradiated by a flood gun through a mask with a 55-μm periodicity (35 μm diameter, 20 μm distance). (c) Tapping-mode AFM image of PNIPAM brushes with a 2.5-μm periodicity (1.8 μm diameter and 0.7 μm distance) in air and corresponding height profile. (Reproduced with permission from [80]. © 2007, American Chemical Society.)

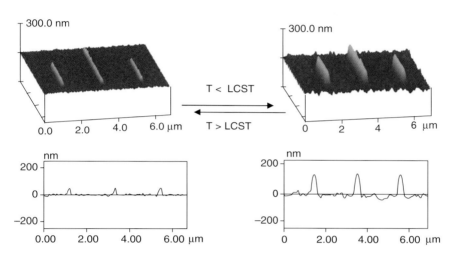

Figure 5.7 AFM images of 70-nm line-width PNIPAM patterns obtained in water, and the corresponding height profile at room temperature and 40 °C, respectively. (Reproduced with permission from [80]. © 2007, American Chemical Society.)

culture experiments have demonstrated that fibroblast cells could adhere to the different-shaped PNIPAM patterns, but the surrounding PEG2000 layer is very effective in preventing cells from attaching and spreading owing to increased hydrophilicity as shown in Figure 5.8 [84, 85]. In other words, the cell body can spread across the interspacing nonadhesive areas (i.e., PEG2000 layer) of the substrate and elongate from one small adhesive dot to another. These results prove that cells can adhere, spread, and proliferate on the patterned PNIPAM film at 37 °C (above the LCST). On the other hand, the study has also shown that a single cell can adhere to the multiple, closely spaced PNIPAM dot patterns, but cannot spread across two dot stripes when the interspaces between two dot stripes (i.e., nonadhesive areas or PEG2000 layer) are too large. With spatial control over where a cell can adhere and where it cannot, we can separate changes in a cell's fate because of the changes of the adhered cell shapes [86, 87]. This point is very important because the physically extended degree of a cell can conveniently determine the proliferation or death of a cell. This fabrication approach allows creating

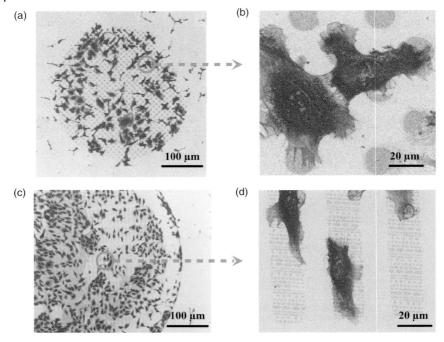

Figure 5.8 SEM images of fibroblasts spread on multiple PNIPAM circle patterns surrounded with PEG2000: (a) circle patterns (diameter: 15 μm; space: 30 μm) and (c) mixed patterns; (b) and (d) are magnified images of (a) and (c), respectively. (Reproduced with permission from [80]. © 2007, American Chemical Society.)

spatially defined polymer patterns, and provides a simple and versatile method to construct complex micro- and nanopatterned polymer brushes with spatial and topographic control in a single step.

5.3.2
Fabrication of Complex Polymer Brush Gradients

With regard to the interactions of biological species with polymer surfaces, one important problem in studies using different kinds of polymer surfaces is that the surfaces are heterogeneous, both chemically and physically (different surface chemistry, charge, roughness, rigidity, crystallinity, etc.), which may result in considerable variation in experimental results. Another methodological problem is that the evaluation of the behavior of biological species on a polymer surface is often tedious because a large number of samples must be prepared to cover the range of a desired variable [88–91]. Thus, there is the strong possibility of methodological error. Many recent studies have focused on the preparation of surfaces with a continuously varying chemical composition along one dimension.

In such so-called "gradient surfaces" the gradually varying chemical composition on the surface produces gradients in wettability, thickness, dielectric constant, or other physicochemical properties. Substrates thus formed find uses in applications including selective adsorption, gradient templating, controlled motion of liquid droplets, particle sorting, and so on. In addition to their broad range of applications, such gradient surfaces are of particular interest for basic studies of the interactions between biological species and surfaces since the effect of a selected property can be examined in a single experiment on one surface. Some groups have reported the preparation of wettability gradient surfaces and their uses in studying interaction phenomena of biological species, including proteins, cells, and enzymes [92–94]. Unfortunately, current preparation methods are troublesome and, in particular, cannot produce more complex three-dimensional polymer structures at a nanometer scale. In the previous section we described an example for the possible applications of electron beam-modified nitro-functionalized SAMs, that is, through the locally confined growth of topographic polymer brushes by covalent attachment of initiator units to the terminal amino groups of nanostructures written into a SAM, brushes of common organic polymers can be prepared by atom transfer free radical polymerization of unsaturated monomers. In the following section, we will introduce our recent work showing that a variety of three-dimensional complex polymer nanostructures can be fabricated by using this combination of chemical lithography with ATRP in a single step [95]. Briefly, by variation of the electron beam dose, the amount of terminal nitro groups converted to amino groups can be controlled and thus the density of polymerization initiator attached in the following step can be varied. After attachment of initiator units, arbitrary three-dimensional structures can be formed in a "one-pot" polymerization (Figure 5.9). Our studies have shown that the brush height increases with the intensity of electron exposure up to a dose of 40 mC cm^{-2}. The height of PNIPAM brushes is a function of electron dose and can be described by:

$$H = D_0 \left[1 - \exp^{-(x/18.2)} \right]$$

where H is the brush height (nm), x is the applied area electron dose (mC cm^{-2}), and D_0 is the measured (final) brush height corresponding to the maximum concentration of amine groups on the surface. It has to be pointed out that the absolute value of D_0 depends on the polymerization conditions, such as polymerization time and the concentration of monomer and catalyst. The dependence of PNIPAM brush height on electron dose is correlated with the efficiency of terminal nitro group reduction and thus initiator density. The conversion of the nitro to amino groups in NBT SAMs has previously been studied by XPS. The surface density of the amino group increases with the increase of electron dose and saturates (for 3.0-keV electrons) at a dose of about 40 mC cm^{-2}. In our experiments the polymerization conditions are identical for all experiments and we assume that the degree of polymerization is independent of initiator density. When the distance between the initiator units is much larger than the radius of gyration of the resulting polymer, the polymer brushes will first fill the lateral space during the

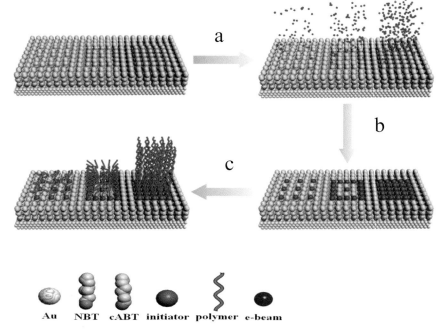

Figure 5.9 Representation of the preparation process of different height polymer brush patterns by chemical lithography and SI-ATRP. (a) A NBT SAM was exposed to the electron beam, resulting in cross-linking of the SAM and conversion of the terminal nitro groups to amino groups. (b) The binding of bromoisobutyryl bromide to the amino sites results in a SAM that bears the surface initiator. (c) Exposure to NIPAM and radical polymerization results in a polymer brush layer at the irradiated regions. cABT, cross-linked 4′-amino-1,1′-biphenyl-4-thiol. (Reproduced with permission from [95]. © 2007, Wiley-VCH.)

polymerization before growing in the third dimension. Only in the case of very high grafting densities will the polymer brushes grow predominantly in the third dimension (Figure 5.10). In other words, gradients of the surface initiator concentration can be produced using a lateral gradient of the electron dose and, in the following step, polymeric thickness gradients can be produced.

Chemical lithography requires only three steps to prepare a topographic polymer brush pattern on a gold surface – formation of a SAM, electronic beam exposure, and polymerization on the activated areas. Thus, the combination of chemical lithography with SI-ATRP can also be used to fabricate complex three-dimensional gradient structures of stimulus-responsive polymer brushes made, for example, from PNIPAM with controlled size, shape, position, and thickness on a gold substrate. Circular PNIPAM structures of 1 µm diameter and different heights have been prepared, and displayed the height of a polymer gradient structure prepared

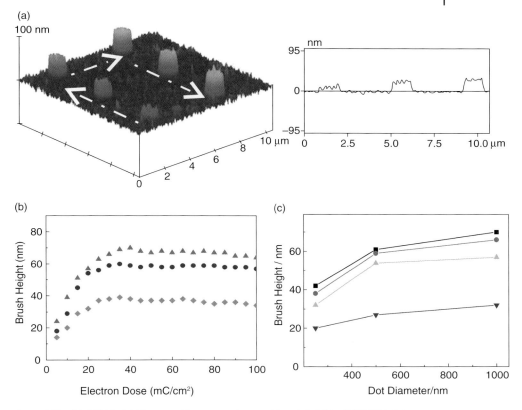

Figure 5.10 (a) AFM height images of PNIPAM brushes obtained after writing in a NBT SAM with different electron doses (1 μm diameter). The colors red, blue, and green represent the brush heights of 0–25, 25–50, and 50–75 nm, respectively. The yellow arrows represent the increase of electron dose. (b) The relation between the electron dose and brush height of PNIPAM dot patterns (▲ = 1000; • = 500; ♦ = 250 nm). (c) Brush height as a function of dot diameter etched by using different electron doses (▲ = 10; ▼ = 20; ♦ = 30; ■ = 40 mC cm^{-2}). (Reproduced with permission from [95]. © 2007, Wiley-VCH.)

with increasing electron doses (Figure 5.11). Furthermore, polymer brush patterns with different widths ranging from 1 μm to 100 nm have been prepared on the same substrate and the enlarged AFM images show that when the polymerization time was extended to 90 min, the brush height approximately increased 1.35 times as compared to those in Figure 5.1 with the same structure and size. This demonstrates that the size, shape, and height of topographic polymer structures can be controlled by electron exposure and polymerization time.

The PNIPAM brush height depends on the density of the initiator (controlled by the electron beam dose in the chemical lithography process) and thus allows

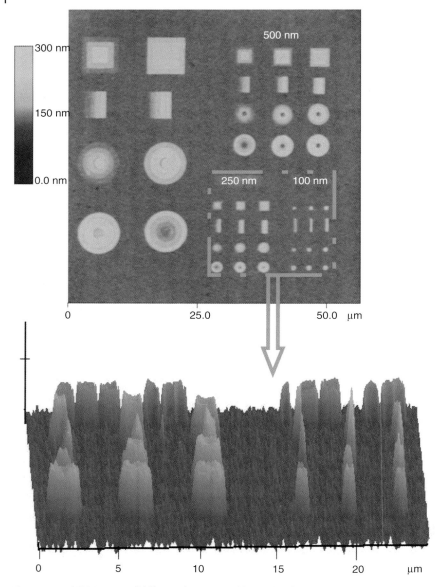

Figure 5.11 AFM images of different diameter and line-width for PNIPAM brush structures of 1000, 500, 250, and 100 nm, and the corresponding magnified micrograph. (Reproduced with permission from [95]. © 2007, Wiley-VCH.)

the preparation of spatially defined polymer patterns with varying heights. The swelling behavior of these stimulus-responsive polymer patterns in water is strongly temperature-dependent and swelling largely follows what is known for bulk PNIPAM hydrogels. Thus, unique PNIPAM surface topographies can be created by controlled variation of electron doses and polymerization conditions. Additionally, below the LCST, the Young modulus E was determined from force–distance curves on a swollen pattern and the corresponding curve section was fitted with the theoretical curve based on the Hertz model. Our studies have shown that the increase of E correlates with the compression of the brush and the closer distance of the AFM tip to the subjacent hard substrate.

In an effort to overcome the need for custom-synthesized aromatic SAM resists, Ballav et al. used aliphatic dodecanethiol [96]. After exposure to the electron beam, the patterns were displaced with 11-aminoundecanethiol (AUT). This resulted in amino-terminated AUT templates with a size of around 50 nm in a background of methyl-terminated DDT. An ATRP bromoinitiator was then attached to the patterned AUT templates for the SIP of PNIPAM brushes. Therefore, chemical lithography has opened a new window of opportunities for the creation of polymer brush nanostructures with high resolution and fidelity. Chemical lithography offers the capability to build at both the micro- and nanoscale while imparting great control over the grafting density and height of the polymer brushes. Furthermore, the chemical modification of SAMs on surfaces provides templates for the fabrication of a wide variety of "smart" polymeric nanostructures.

5.4
Conclusions and Perspectives

Fundamental understanding of the interactions of biomolecules binding to substrates is essential for developing workable technologies for life science. The new capabilities to study and control processes on the nanometer scale are emerging as valuable assets in both fundamental and applied research. Potential applications include the development of a new generation of chemical and biosensors, biochips, and molecular electronic devices. We anticipate that progress of nanotechnology and life sciences will greatly accelerate the development of biomimetic surface research for drug delivery, biomolecule-based devices, or biosensing. In this chapter, we have tried to provide an insight into the tremendous versatility of several nanofabrication methods for protein or lipid nanopatterning. In addition, polymer brush nanopatterning or complex gradients can be conveniently fabricated by using the combination of chemical lithography and living radical polymerization, which can be used to mimic the extracellular matrix, and systematically investigate the various types of biological species and polymer interactions in terms of the surface wettability of the polymer. Importantly, the complex gradient of polymer brushes could be identified where the number of adhering proteins and cells, and their corresponding spread areas, were maximal. This means that certain surface properties are effectively compared at the same time and on the same surface. Therefore, polymer gradient surfaces offer potential not only as models for molecular recognition and interactions in biological systems, and for cell motility and diagnostic studies, but also for practical applications such as cell separation, drug delivery systems, and sensors in biotechnology.

References

1 Niewmeyer, C.M. and Mirkin, C.A. (eds) (2004) *Nanobiotechnology: Concepts, Applications and Perspectives*, Wiley-VCH Verlag GmbH, Weinheim.
2 Kumar, C. (ed.) (2007) *Nanodevices for the Life Sciences*, Wiley-VCH Verlag GmbH, Weinheim.
3 He, Q., Cui, Y., and Li, J.B. (2009) Molecular assembly and application of biomimetic microcapsules. *Chem. Soc. Rev.*, **38**, 2292–2303.
4 Bao, G. and Suresh, S. (2003) Cell and molecular mechanics of biological materials. *Nat. Mater.*, **2**, 715–725.
5 Ratner, B.D. and Bryant, S.J. (2004) Biomaterials: where we have been and where we are going. *Annu. Rev. Biomed. Eng.*, **6**, 41–75.
6 Sarikaya, M., Tamerler, C., Yen, A.K.-Y., Schulten, K., and Baneyx, F. (2003) Molecular biomimetics: nanotechnology through biology. *Nat. Mater.*, **2**, 577–585.
7 Sanchez, C., Arribart, H., Madeleine, M., and Guille, G. (2005) Biomimetism and bioinspiration as tools for the design of innovative materials and systems. *Nat. Mater.*, **4**, 277–288.
8 Scouten, W.H., Luong, J.H.T., and Brown, R.S. (1995) Enzyme or protein immobilization techniques for applications in biosensor design. *Trends Biotechnol.*, **13**, 178–185.
9 Whitesides, G.M., Ostuni, E., Takayama, S., Jiang, X., and Ingber, D.E. (2001) Soft lithography in biology and biochemistry. *Annu. Rev. Biomed. Eng.*, **3**, 335–373.
10 Zhang, S.G., Yan, L., Altman, M., Lassle, M., Nugent, H., Frankel, F., Lauffenburger, D.A., and Whitesides, G.M. (1999) Biological surface engineering: a simple system for cell pattern formation. *Biomaterials*, **20**, 1213–1220.
11 Bernard, A., Renault, J.P., Michel, B., Bosshard, H.R., and Delamarche, E. (2000) Microcontact printing of proteins. *Adv. Mater.*, **12**, 1067–1070.
12 Zhang, X.M., He, Q., Cui, Y., Duan, L., and Li, J.B. (2006) Human serum albumin supported lipid patterns for the targeted recognition of microspheres coated by membrane based on ssDNA hybridization. *Biochem. Biophys. Res. Comm.*, **349**, 920–924.
13 Dontha, N., Nowall, W.B., and Kuhr, W.G. (1997) Generation of biotin/avidin/enzyme nanostructures with maskless photolithography. *Anal.Chem.*, **69**, 2619–2625.
14 Delamarche, E., Bernard, A., Schmid, H., Bietsch, A., Michel, B., and Biebuyck, H. (1998) Microfluidic networks for chemical patterning of substrate: design and application to bioassays. *J. Am. Chem. Soc.*, **120**, 500–508.
15 Dameron, A.A., Hampton, J.R., Smith, R.K., Mullen, T.J., Gillmor, S.D., and Weiss, P.S. (2005) Microdisplacement printing. *Nano Lett.*, **5**, 1834–1837.
16 Geissler, M., McLellan, J.M., Chen, J., and Xia, Y. (2005) Side-by-side patterning of multiple alkanethiolate monolayers on gold by edge-spreading lithography. *Angew. Chem. Int. Ed.*, **44**, 3596–3600.
17 Chou, S.Y., Krauss, P.R., and Renstrom, P.J. (1996) Imprint lithography with 25-nanometer resolution. *Science*, **272**, 85–87.
18 Bietsch, A., Hegner, M., Lang, H.P., and Gerber, C. (2004) Inkjet deposition of alkanethiolate monolayers and DNA oligonucleotides on gold: evaluation of spot uniformity by wet etching. *Langmuir*, **20**, 5119–5122.
19 Glass, R., Arnold, M., Blümmel, J., Küller, A., Möller, M., and Spatz, J.P.

(2003) Micro-nanostructured interfaces fabricated by the use of inorganic block copolymer micellar monolayers as negative resist for electron-beam lithography. *Adv. Funct. Mater.*, **13**, 569–575.
20. Smith, R.K., Lewis, P.A., and Weiss, P.S. (2004) Patterning self-assembled monolayers. *Prog. Surf. Sci.*, **75**, 1–68.
21. Attwood, D., Anderson, E., Denbeaux, G., Goldberg, K., Naulleau, P., and Schneider, G. (2002) Soft x-ray microscopy and EUV lithography: an update on imaging at 20–40 nm spatial resolution. *AIP Conf. Proc.*, **641**, 461–468.
22. Klauser, R., Hong, I.H., Wang, S.C., Zharnikov, M., Paul, A., Gölzhäuser, A., Terfort, A., and Chuang, T.J. (2003) Imaging and Patterning of monomolecular resists by zone-plate-focused X-ray microprobe. *J. Phy. Chem. B*, **107**, 13133–13142.
23. Giannuzzi, L.A. and Stevie, F.A. (eds) (2005) *Introduction to Focused Ion Beams*, Springer, New York.
24. Li, M., Chen, L., and Chou, S.Y. (2001) Direct three-dimensional patterning using nanoimprint lithography. *Appl. Phys. Lett.*, **78**, 3322–3324.
25. Ginger, D.S., Zhang, H., and Mirkin, C.A. (2004) The evolution of dip-pen nanolithography. *Angew. Chem. Int. Ed.*, **43**, 30–45.
26. Salaita, K., Wang, Y., and Mirkin, C.A. (2007) Applications of dip-pen nanolithography. *Nat. Nanotechnol.*, **2**, 145–155.
27. Liu, M., Amro, N.A., and Liu, G.-Y. (2008) Nanografting for surface physical chemistry. *Annu. Rev. Phys. Chem.*, **59**, 367–386.
28. Lahiri, J., Ostuni, E., and Whitesides, G.M. (1999) Patterning ligands on reactive SAMs by microcontact printing. *Langmuir*, **15**, 2055–2060.
29. Wong, S.S. (1991) *Chemsitry of Protein Conjugation and Cross-Linking*, CRC Press, Boca Raton, FL.
30. Hermanson, G.T. (1996) *Bioconjugate Techniques*, Academic Press, San Diego, CA.
31. Zhang, X.M., He, Q., Yan, X.H., Boullanger, P., and Li, J.B. (2007) Glycolipid patterns supported by human serum albumin for *E. coli* recognition. *Biochem. Biophys. Res. Commun.*, **358**, 424–428.
32. Lu, G., An, Z.H., and Li, J.B. (2004) Biogenic capsules made of proteins and lipids. *Biochem. Biophys. Res. Commun.*, **315**, 224–227.
33. Lu, G., An, Z.H., Tao, C., and Li, J.B. (2004) Microcapsule assembly of human serum albumin at the liquid/liquid interface by the pendent drop technique. *Langmuir*, **20**, 8401–8403.
34. An, Z.H., Lu, G., Möhwald, H., and Li, J.B. (2004) Self-assembly of human serum albumin (HSA) and L-alpha-dimyristoylphosphatidic acid (DMPA) microcapsules for controlled drug release. *Chem. Eur. J.*, **10**, 5848–5852.
35. An, Z.H., Tao, C., Lu, G., Möhwald, H., Zheng, S.P., Cui, Y., and Li, J.B. (2005) Fabrication and characterization of human serum albumin and L-alpha-dimyristoylphosphatidic acid microcapsules based on template technique. *Chem. Mater.*, **17**, 2514–2519.
36. An, Z.H., Möhwald, H., and Li, J.B. (2006) pH controlled permeability of lipid/protein biomimetic microcapsules. *Biomacromolecules*, **7**, 580–585.
37. Hakomori, S. and Handa, K. (2002) Glycosphingolipid-dependent cross-talk between glycosynapses interfacing tumor cells with their host cells: essential basis to define tumor malignancy. *FEBS Lett.*, **531**, 88–92.
38. Smith, A.E. and Helenius, A. (2004) How viruses enter animal cells. *Science*, **304**, 237–242.
39. Fazio, F., Bryan, M.C., Blixt, O., Paulson, J.C., and Wong, C.H. (2002) Synthesis of sugar arrays in microtiter plate. *J. Am. Chem. Soc.*, **124**, 14397–14402.
40. Fukui, S., Feizi, T., Galustian, C., Lawson, A.M., and Chai, W. (2002) Oligosaccharide microarrays for high-throughput detection and specificity assignments of carbohydrate–protein interactions. *Nat. Biotechnol.*, **20**, 1011–1017.
41. Park, S., and Shin, I. (2002) Fabrication of carbohydrate chips for studying protein–carbohydrate interactions. *Angew. Chem. Int. Ed.*, **41**, 3180–3182.

42 Houseman, B.T., and Mrksich, M. (2002) Carbohydrate arrays for the evaluation of protein binding and enzymatic modification. *Chem. Biol.*, **9**, 443–454.

43 Disney, M.D., Zhang, J., Swager, T.M., and Seeberger, P.H. (2004) Detection of bacteria with carbohydrate-functionalized fluorescent polymers. *J. Am. Chem. Soc.*, **126**, 13343–13346.

44 Boullanger, P., Sancho-Camborieux, M.R., Bouchu, M.N., Marron-Brignone, L., Morelis, R.M., and Coulet, P.R. (1997) Synthesis and interfacial behavior of three homologous glycero neoglycolipids with various chain lengths. *Chem. Phys. Lipids*, **90**, 63–74.

45 Case, M.A., Mclendon, G.L., Hu, Y., Vanderlick, T.K., and Scoles, G. (2003) Using nanografting to achieve directed assembly of *de novo* designed metalloproteins on gold. *Nano Lett.*, **3**, 425–429.

46 Wilson, D.L., Martin, R., Hong, S., Cronin-Golomb, M., Mirkin, C.A., and Kaplan, D.L. (2001) Surface organization and nanopatterning of collagen by dip-pen nanolithography. *Proc. Natl. Acad. Sci. USA*, **98**, 13660–13664.

47 Wadu-Mesthrige, K., Xu, S., Amro, N.A., and Liu, G.Y. (1999) Fabrication and imaging of nanometer-sized protein patterns. *Langmuir*, **15**, 8580–8583.

48 Jang, C.H., Stevens, B.D., Philips, R., Calter, M.A., and Drucker, W.A. (2003) A strategy for the sequential patterning of proteins: catalytically active multiprotein nanofabrication. *Nano Lett.*, **3**, 691–694.

49 Nuzzo, R.G. and Allara, D.L. (1983) Adsorption of bifunctional organic disulfides on gold surfaces. *J. Am. Chem. Soc.*, **105**, 4481–4483.

50 Ulman, A. (1996) Formation and structure of self-assembled monolayers. *Chem. Rev.*, **96**, 1533–1544.

51 Geyer, W., Stadler, V., Eck, W., Zharnikov, M., Gölzhäuser, A., and Grunze, M. (1999) Electron-induced cross-linking of aromatic self-assembled monolayers: negative resists for nanolithography. *Apple. Phys. Lett.*, **75**, 2401–2403.

52 Eck, W., Stadler, V., Geyer, W., Zharnikov, M., Gölzhäuser, A., and Grunze, M. (2000) Generation of surface amino groups on aromatic self-assembled monolayers by low energy electron beams – a first step towards chemical lithography. *Adv. Mater.*, **12**, 805–808.

53 Gölzhäuser, A., Eck, W., Geyer, W., Stadler, V., Weimann, T., Hinze, P., and Grunze, M. (2001) Chemical nanolithography with electron beams. *Adv. Mater.*, **13**, 806–809.

54 Turchanin, A., Tinazli, A., El-Desawy, M., Grossmann, H., Schnietz, M., Solak, H.H., Tampe, R., and Gölzhäuser, A. (2008) Molecular self-assembly, chemical lithography, and biochemical tweezers: a path for the fabrication of functional nanometer-scale protein arrays. *Adv. Mater.*, **20**, 471–477.

55 Biebricher, A., Paulb, A., Tinnefeld, P., Gölzhäuser, A., and Sauer, M. (2004) Controlled three-dimensional immobilization of biomolecules on chemically patterned surfaces. *J. Biotechnol.*, **112**, 97–107.

56 Sackmann, E. (1996) Supported membranes: scientific and practical applications. *Science*, **271**, 43–48.

57 McConnell, H.M., Watts, T.H., Weis, R.M., and Brian, A.A. (1986) Supported planar membranes in studies of cell–cell recognition in the immune system. *Biochim. Biophys. Acta*, **271**, 95–106.

58 Tanaka, M. and Sackmann, E. (2005) Polymer-supported membranes as models of the cell surface. *Nature*, **437**, 656–663.

59 Richter, R.P., Berat, R., and Brisson, A.R. (2006) Formation of solid-supported lipid bilayers: an integrated view. *Langmuir*, **22**, 3497–3505.

60 Groves, J.T. and Boxer, S.G. (2002) Micropattern formation in supported lipid membranes. *Acc. Chem. Res.*, **35**, 149–157.

61 Groves, J.T., Ulman, N., and Boxer, S.G. (1997) Micropatterning fluid lipid bilayers on solid supports. *Science*, **275**, 651–653.

62 Hovis, J.S. and Boxer, S.G. (2000) Patterning barriers to lateral diffusion in supported lipid bilayer membranes by blotting and stamping. *Langmuir*, **16**, 894–897.

63 He, Q., Tian, Y., Küller, A., Grunze, M., Kuller, A., Gölzhäuser, A., and Li, J.B. (2006) Self-assembled molecular pattern by chemical lithography and interfacial chemical reactions. *J. Nanosci. Nanotechnol.*, **6**, 1838–1841.

64 Wang, X.L., He, Q., Zheng, S.P., Brezesinski, G., Möhwald, H., and Li, J.B. (2004) Structural changes of phospholipid monolayers caused by coupling of human serum albumin: a GIXD study at the air/water interface. *J. Phys. Chem. B*, **108**, 14171–14177.

65 Möwhald, H. (1990) Phospholipid and phospholipid-protein monolayers at the air/water interface. *Annu. Rev. Phy. Chem.*, **41**, 441–476.

66 Ratner, B.D., Hoffman, A.S., Schoen, F.J., and Lemons, J.E. (2004) *Biomaterials Science: An Introduction to Materials in Medicine*, Academic Press, New York.

67 Geoghegan, M. and Krausch, G. (2003) Wetting at polymer surfaces and interfaces. *Prog. Polym. Sci.*, **28**, 261–302.

68 Kim, M.S., Khang, G., and Lee, H.B. (2008) Gradient polymer surfaces for biomedical applications. *Prog. Polym. Sci.*, **33**, 138–164.

69 Curtis, A. and Wilkinson, C. (1997) Topographical control of cells. *Biomaterials*, **18**, 1573–1583.

70 Kim, S.K., Teixeira, A.I., Nealey, P.F., Wendt, A.E., and Abbott, N.L. (2002) Fabrication of polymeric substrates with well-defined nanometer-scale topography and tailored surface chemistry. *Adv. Mater.*, **14**, 1468–1472.

71 Piner, R.D., Zhu, J., Xu, F., Hong, S., and Mirkin, C.A. (1999) "Dip-pen" nanolithography. *Science*, **283**, 661–663.

72 Xia, Y. and Whitesides, G.M. (1998) Soft lithography. *Angew. Chem. Int. Ed.*, **37**, 550–575.

73 Ito, T. and Okazaki, S. (2000) Pushing the limits of lithography. *Nature*, **406**, 1027–1031.

74 Ducker, R., Garcia, A., Zhang, J.M., Chen, T., and Zauscher, S. (2008) Polymeric and biomacromolecular brush nanostructures: progress in synthesis, patterning and characterization. *Soft Matter*, **4**, 1774–1786.

75 Zhao, B. and Brittain, W.J. (2000) Polymer brushes: surface-immobilized macromolecules. *Prog. Polym. Sci.*, **25**, 677–710.

76 Advincula, R.C., Brittain, W.J., Caster, K.C., and Ruehe, J. (eds) (2004) *Polymer Brushes*, Wiley-VCH Verlag GmbH, Weinheim.

77 Edmondson, S., Osborne, V.L., and Huck, W.T.S. (2004) Polymer brushes via surface-initiated polymerizations. *Chem. Soc. Rev.*, **33**, 14–22.

78 Senaratne, W., Andruzzi, L., and Ober, C.K. (2005) Self-assembled monolayers and polymer brushes in biotechnology: current applications and future perspectives. *Biomacromolecules*, **6**, 2427–2448.

79 Matyjaszewski, K. and Xia, J.H. (2001) Atom transfer radical polymerization. *Chem. Rev.*, **101**, 2921–2990.

80 He, Q., Kueller, A., Grunze, M., and Li, J.B. (2007) Fabrication of thermosensitive polymer nanopatterns through chemical lithography and atom transfer radical polymerization. *Langmuir*, **23**, 3981–3987.

81 Li, D.X., He, Q., and Li, J.B. (2009) Smart core/shell nanocomposites: intelligent polymers modified gold nanoparticles. *Adv. Colloid Interface Sci.*, **149**, 28–38.

82 van Wachem, P.B., Beugeling, T., Feijen, J., Bantjes, A., Detmaers, J.P., and van Aken, W.G. (1985) Interaction of cultured human endothelial cells with polymeric surfaces of different wettabilities. *Biomaterials*, **6**, 403–408.

83 Gristina, A.G. (1987) Biomaterial-centered infection–microbial adhesion versus tissue integration. *Science*, **237**, 1588–1595.

84 Gotoh, Y., Tsukada, M., Minoura, N., and Imai, Y. (1997) Synthesis of poly(ethylene glycol)–silk fibroin conjugates and surface interaction between L-929 cells and the conjugates. *Biomaterials*, **18**, 267–271.

85 Wang, P., Tan, K.L., Kang, E.T., and Neoh, K.G. (2001) Surface functionalization of low density polyethylene films with grafted poly(ethylene glycol) derivatives. *J. Mater. Chem.*, **11**, 2951–2957.

86 Chen, C.S., Mrksich, M., Huang, S., Whitesides, G.M., and Ingber, D.E.

(1997) Geometric control of cell life and death. *Science*, **276**, 1425–1428.
87 Sniadecki, N.J., Desai, R.A., Ruiz, S.A., and Chen, C.S. (2006) Nanotechnology for cell–substrate interactions. *Ann. Biomed. Eng.*, **34**, 59–74.
88 Ruardy, T.G., Schakenraad, J.M., van der Mei, H.C., and Busscher, H.J. (1997) Preparation and characterization of chemical gradient surfaces and their application for the study of cellular interaction phenomena. *Surf. Sci. Rep.*, **29**, 3–30.
89 Ruardy, T.G., Schakenraad, J.M., van der Mei, H.C., and Busscher, H.J. (1995) Adhesion and spreading of human skin fibroblasts on physicochemically characterized gradient surfaces. *J. Biomed. Mater. Res.*, **29**, 1415–1423.
90 Pitt, W.G. (1989) Fabrication of a continuous wettability gradient by radio-frequency plasma discharge. *J. Colloid Interface Sci.*, **133**, 223–227.
91 Bhat, R.R., Tomlinson, M.R., Wu, T., and Genzer, J. (2006) Surface-grafted polymer gradients: formation, characterization, and applications. *Adv. Polym. Sci.*, **198**, 51–124.
92 Efimenko, K. and Genzer, J. (2001) How to prepare tunable planar molecular chemical gradients. *Adv. Mater.*, **13**, 1560–1563.
93 Bhat, R.R., Chaney, B.N., Rowley, J., Liebmann-Vinson, A., and Genzer, J. (2005) Tailoring cell adhesion using surface-grafted polymer gradient assemblies. *Adv. Mater.*, **17**, 2802–2807.
94 Wu, T., Efimenko, K., and Genzer, J. (2002) Combinatorial study of the mushroom-to-brush crossover in surface anchored polyacrylamide. *J. Am. Chem. Soc.*, **124**, 9394–9395.
95 He, Q., Kueller, A., Schilp, S., Leisten, F., Kolb, H., Grunze, M., and Li, J.B. (2007) Fabrication of controlled thermosensitive polymer nanopatterns with one-pot polymerization through chemical lithography. *Small*, **3**, 1860–1865.
96 Ballav, N., Schilp, S., and Zharnikov, M. (2008) Electron-beam chemical lithography with aliphatic self-assembled monolayers. *Angew. Chem. Int. Ed.*, **47**, 1421–1424.

6
Peptide-Based Biomimetic Materials

6.1
Introduction

Molecular assembly of biological and biomimetic building blocks is an emerging hot field in which hybrid technologies are developed by using the tools of molecular biology and nanotechnology [1–3]. By taking lessons from nature or imitating certain aspects of biological systems, a variety of biological and biomimetic materials have been constructed via a "bottom-up" approach – molecular self-assembly [4–8]. This type of assembly is ubiquitous in nature, such as in the formation of biological membranes upon self-assembly of phospholipids, DNA double-helix formation through specific hydrogen bonding interactions, protein microtubules and microfilaments as functional units for intracellular interplay, as well as the formation of amyloid fibrils in a variety of neurological diseases. Clearly, self-assembly plays a vital role in many biological systems, either to achieve biological function or as part of a pathogenic process. All biomolecules are able to interact each other and undergo self-assembly into ordered structures. In a certain sense, these molecules can mimic biological systems and, in turn, provide new insights into the original biological function.

The ordered organization of these biological and biomimetic building blocks into defined structures is done on the basis of an association of some weak noncovalent bonds, notably including hydrogen bonds, electrostatic interactions, π–π stacking, hydrophobic forces, nonspecific van der Waals forces, and chiral dipole–dipole interactions [5]. Although these forces are relatively weak individually, when combined together, they can govern the assembly of molecular building blocks into superior and ordered structures. they are also weak compared to thermal forces and of similar magnitude, thus enabling variations of structures and properties by small variation of parameters.

Nanostructures fabricated from biomolecules are attracting increasing attention owing to their biocompatibility, specific molecular recognition ability, simple chemical and biological modifiability, and easy availability for bottom-up fabrication. Many biomolecules, such as lipids, nucleic acids, proteins, and peptides, can interact and self-assemble into highly ordered supramolecular architectures.

Molecular Assembly of Biomimetic Systems. Junbai Li, Qiang He, and Xuehai Yan
© 2011 WILEY-VCH Verlag GmbH & Co. KGaA, Weinheim
ISBN: 978-3-527-32542-9

Among them, peptides composed of several to dozens of amino acids are a class of versatile building blocks for this purpose [5, 9]. In some cases, they can also serve as analogs of proteins, offering an alternative model for understanding the self-assembly and functions of proteins. Furthermore, the self-assembly nature of designed or extracted peptide building blocks enables them to be readily manipulated to well-defined nanostructures with various functions. Over the past decades, researchers have made significant progress in this field. A number of peptide-based building blocks, including cyclic peptides [10], dendritic peptides [11], peptide amphiphiles [12, 13], surfactant-like oligopeptides [14, 15], polypeptides [16, 17], and aromatic dipeptides [9, 18], have been designed and developed for the creation of functional supramolecular architectures, and the exploration of their possible applications in biology and in nanotechnology.

In this chapter, we first focus on the fabrication of peptide-based nanostructural materials from synthetic building blocks such as lipopeptides, polypeptides, amphiphilic peptides, and, particularly, diphenylalanine-based peptides derived from Alzheimer's β-amyloid (Aβ) polypeptide. In addition, we present the experimental results and the progress in the integration of peptide biomaterials with functional inorganic components for creating multifunctional materials. We then discuss the potential applications of such assembled peptide-based materials in biological and nonbiological areas, including tissue engineering, gene or drug delivery, bioimaging and biosensors, as well as functional templates for nanofabrication.

6.2
Peptides as Building Blocks for the Bottom-up Fabrication of Various Nanostructures

6.2.1
Aromatic Dipeptides

In a variety of eminent short peptide building blocks, diphenylalanine peptide (L-Phe–L-Phe, FF) is considered one of the simplest and most effective for bottom-up fabrication. This finding is based on the study of the fibrillization mechanism of Alzheimer's Aβ polypeptide. It is known that the formation of amyloid plaques and neurofibrillary tangles are thought to cause the degradation of neurons ("nerve cells") in the brain and subsequent symptoms of Alzheimer's disease. Amyloid, a general term for protein fragments, refers to protein deposits resembling those first observed for starch (amyloid originally meant starch-like) [19]. In Alzheimer's disease, the protein fragments cannot be normally broken down and cleared, and thus accumulate to form hard and insoluble plaques. The Aβ peptide, a major component of senile plaques that is diagnostic of Alzheimer's disease, is now widely accepted as the causative agent [20]. Such a pathogenic process concerning amyloid fibrils provides important information on intermolecular interactions and thus gives us considerable inspiration for devising new types of self-assembling

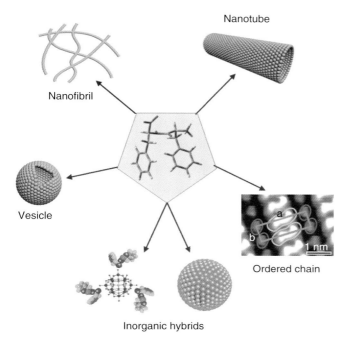

Figure 6.1 Schematic representation of various nanostructures formed by self-assembly of FF-based peptide building blocks.

building blocks. A known example is FF, which is extracted from Alzheimer's Aβ polypeptide as the core recognition motif for molecular self-assembly [21]. FF-based building blocks have been assembled into various nanostructures such as nanotubes, spherical vesicles, nanofibrils and ribbons, nanowires and ordered molecular chains, and so on (Figure 6.1).

6.2.1.1 Nanotubes, Nanotube Arrays, and Vesicles

Proteins and peptides are very attractive for the construction of tubular nanostructures, and rapid progress has been made in the development of peptide-based nanotubes [22]. The simplest peptide building block for self-assembly is generally regarded as the FF dipeptide. It has been demonstrated that FF can self-assemble into well-ordered tubular structures with a persistent length by a combination of hydrogen bonding and π–π stacking of aromatic residues (Figure 6.2a) [21, 23]. The ordered molecular organization of FF nanotubes (FNTs) is attributed to a striking three-dimensional aromatic stacking alignment that serves as a glue between the hydrogen-bonded cylinders of the peptide main-chain [24]. The tubular nanostructures are considerably rigid entities with a high Young's modulus of around 19 GPa [25] (27 GPa in another study, where a bending-beam model was applied to Atomic force microscopy (AFM) images of FNTs to obtain the Young's modulus [26]).

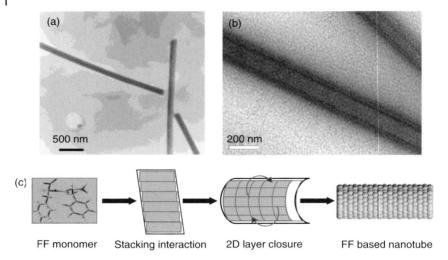

Figure 6.2 TEM image of FNTs (a) and CDPNTs (b). (Reproduced in part with permission from [23] and [30]. © 2004, American Chemical Society and © 2007, Wiley.) (c) Proposed schematic illustration of formation of FF-based nanotubes.

Macroscopic organization or alignment for the self-assembled peptide nanotubes is achieved on demand on the surface of solid substrates. The FNTs have been manipulated into well-organized films on various substrates (SiO_2, gold, palladium, alumina, mica, quartz, InP) by self-assembly of FF in an appropriate solvent, such as N-methyl-2-pyrrolidone, which either acts as the disassembly agent or allows FNTs to reorganize in a film-growth form during the solvent evaporation [27]. In addition to the formation of the FNT film itself, silver-incorporated FNT composite films are also obtained by using a similar method combined with inclusion chemistry. A vertically aligned array of FNTs is achieved by axial unidirectional growth of nanotubes [28]. This growth process is concerned with the evaporation-initiated self-assembly process, in which FF solution in 1,1,1,3,3,3-hexafluoro-2-propanol (HFIP) is spread over a siliconized glass substrate; a thin layer composed of nanotube arrays can be formed upon rapid evaporation of HFIP. Individual FNTs can also be well aligned through exposing them to a strong magnetic field [29]. The aromatic moieties of FF play an important role in the alignment of FNTs. The ordered orientation of aromatic rings is mainly responsible for the net anisotropy in this structure. These unique properties make it possible to apply such peptide nanotubes for the fabrication of biocompatible nanodevices.

A cationic dipeptide (H-Phe–Phe-$NH_2 \cdot HCl$) derived from FF was recently reported capable of self-assembling into nanotubes (hereafter referred to as CDPNTs) at physiological pH [30, 31]. Morphological features of the tubular structure were verified by scanning electron microscopy (SEM), transmission electron microscopy (TEM), and AFM studies. A TEM image confirmed the existence of a typical tubular structure with enough contrast to distinguish the inner part and the periphery of the nanotube (Figure 6.2b). The SEM and AFM studies further support the morphological feature observed in the TEM images, showing the three-dimensional topographic structure and the persistent length. The Circular dichroism (CD) signature obtained on the CDPNTs has some similarities with that of α-helical polypeptides, and the observed extrema may correspond to α-helical $\pi \rightarrow \pi^*$ and $n \rightarrow \pi^*$ transitions, respectively. The route to form tubular structures is proposed in Figure 6.2(c). Peptide monomers first stack with each other to form a two-dimensional layer and then allow the closure of such two-dimensional layer along one axial direction, thus leading to the tubular structures.

Excitingly, a spontaneous conversion of nanotubes into spherical vesicle-like structures takes place by diluting the solution containing CDPNTs at pH 7.2. The structural transition of self-assembled materials is able to be commonly observed and achieved on demand. Parameters to control the morphology mainly involve the molecular and solution parameters. For a given building molecule, molecular parameters such as hydrophobic/hydrophilic properties and the architecture are constant. Nevertheless, different morphologies still can be obtained by tuning solution parameters such as the type of solvents and solvent quality, building molecule concentration, pH value, temperature, and so on [32]. The concentration-dependent structural transition in the peptide-based building blocks, such as linear surfactant-like oligopeptides and D-Phe–D-Phe dipeptide, has been confirmed [14, 33]. As shown in Figure 6.3(a and b), the transition between CDPNTs and vesicle-like structures is reversible dependent on the concentration of peptide building blocks, indicating that the concentration plays a critical role in determining the final nanostructure morphology. In other words, the switch on/off between tubular and spherical structures is able to be readily modulated by varying the concentration of peptide building blocks. Joined spheres in a necklace-like structure, forming an intermediate state of the transition of CDPNTs into vesicle-like structures, were observed directly by using TEM and fluorescence microscopy (Figure 6.3c and d). Similar to the self-assembly of FNTs, the force to drive the formation of CDPNTs may be hydrogen bonding and π–π stacking. However, the X-ray diffraction (XRD) pattern of CDPNTs is somewhat different from that of FNTs, indicating that a possible difference exists in the organized mode of cationic dipeptide [31]. This is attributed to the dipeptide charge state and the nature of the counterions eventually presented, which affects the molecular arrangement of the cationic dipeptide to some extent. At higher concentrations of dipeptide, sufficient free energy of association by the intermolecular interactions can be gained, resulting in the tubular nanostructure. The molecular rearrangement of the dipeptide probably occurs with the decrease of the concentration and thus leads to the formation of vesicle-like structures, which minimizes the free energy of the system. Similar to the result from the Monte Carlo simulation that was carried out by Song *et al.* [33], the dipeptide behaves somewhat like a surfactant; the polar groups isolate from the hydrophobic aromatic groups to form bilayers. With increasing dipeptide concentration, the molecular stacking may be spherical bilayer vesicles, unilamellar nanotubes, multilamellar structures, and continuous phases, respectively.

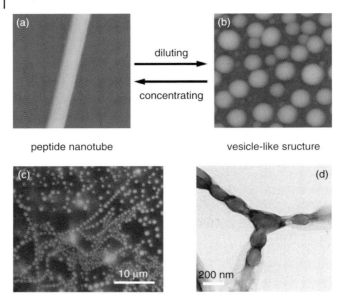

Figure 6.3 Reversible transition between peptide nanotubes and vesicle-like structures: height AFM image of nanotubes (a) and vesicle-like structures (b); fluorescent optical image (c) and TEM image (d) of the joined necklace-like structures composed of spherical vesicles. (Reproduced with permission from [30] and [31]. © 2007, Wiley and © 2008, Wiley.)

According to the multilamellar model of the dipeptide, a general theory model has been proposed to gain further insight into the morphological transition between CDPNTs and vesicle-like structures [31]. To describe the formation of different shapes, the following two-stage mechanism needs to be taken into account [34, 35]. First, monomer molecules are transferred from the isotropic phase ("I-phase") to the outermost face of the aggregate phase ("A-phase") to form a layer of thickness where the layer must gain an internal cohesion energy to compensate for the obvious entropy decrease of the transferred molecules. Additionally, the obtained cohesion energy is mechanically balanced through forming a definite shape that minimizes the total volume and surface free energy. For our system involving dipeptide structural transition, the aggregation at the first stage is readily tuned by the changing peptide concentration, so the gained internal stress of the formed layer also varies with peptide concentration, leading to the shape transition. Following this mechanism, we derive a theory equation as shown below that is defined as the critical tube vesicle concentration (CTVC) [31]:

$$\text{CTVC} = C_A e^{-3\gamma/C_A d k_B T}$$

where C_A is the molecule concentration of the aggregate phase (which is a constant for a given monomer), γ is the tension of solution/aggregate interface, d is the

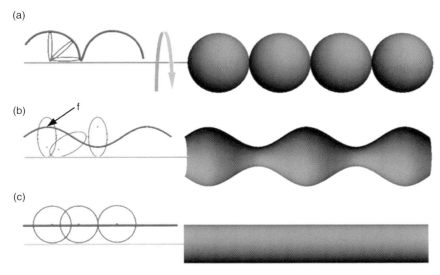

Figure 6.4 Illustration of Delaunay's construction method: (a) sphere, (b) neck-like structure, and (c) tube with constant mean curvature. f is indicative of the focus of an ellipse. (Reprinted with permission from [31]. © 2008, Wiley.)

molecular size, k_B is the Boltzmann constant, and T is the temperature. The CTVC value has a clear physical significance: When $C_S <$ CTVC, nanotubes must transform into spherical vesicle-like structures rapidly, in accordance with the experimental observation in which dipeptide nanotubes can spontaneously change into metastable necklace-like structures and finally into spherical vesicle-like structures. To intuitively comprehend the shape transition, a rotationally symmetric hypersurface can be constructed by Delaunay's method: by rolling a given conical section on a straight line in a plane and then rotating the trace of a focus about the line, one obtains such a surface. In our cases, the conic section is assumed to be an ellipse. A beaded structure is obtained by the Delaunay construction with a thin ellipse section (Figure 6.4a). If the ellipse is very flat, then the beaded structure becomes vesicles of a necklace-like structure (Figure 6.4b). At the other limit, if the ellipse becomes a circle, the resulting surface is a cylindrical tube (Figure 6.4c). All in all, the presented theory model proposes a method to engineer assembling molecules, in order to devise other systems whose morphology could be tuned on demand. Three kinds of free energy (internal cohesion energy, interface energy, and curvature elastic energy) are mainly responsible for this class of shape transition related to monomer concentration. The theory model on CTVC has a critical significance for understanding shape transition in molecular self-assembly.

The introduction of a thiol group to FF dipeptide can change the self-assembly property of the resultant building blocks. For instance, the coupling of cysteine to the FF backbone (i.e., cysteine–diphenylalanine tripeptide, CFF) results in the formation of a spherical vesicle instead of a tubular structure. Likewise, the vesicular structure was also obtained upon chemically reacting a thiol group to the amino terminal of FF. This is ascribed to the energetic contribution provided by the disulfide cross-links, which makes it possible to bend and to close the stacking layer along two axes [23].

6.2.1.2 Nanofibrils and Ribbons

Fibril formation upon self-assembly of peptides and proteins is ubiquitous in biology. In particularly, peptide fibrillization is relevant to a number of diseases, including Alzheimer's disease, Parkinson's disease, Huntington disease, type II diabetes, and prion disorders with the characteristic of deposition of amyloid fibrils in various tissues and organs [19, 36, 37]. On the other hand, nanofibrils self-assembled from natural or *de novo* designed peptides display remarkably potential for applications in biology and nanotechnology [18]. Therefore, peptide nanofibrils constitute one of the most abundant and important naturally occurring self-assembled materials.

A 9-fluorenylmethoxycarbonyl (Fmoc)-protected diphenylalanine (Fmoc-FF) was observed to self-assemble into nanofibrils in water, resulting in the hydrogel being held together by a network of hydrogen bonding and π–π interactions [38, 39]. Such a hydrogel is self-supporting and has a rheological behavior where the storage modulus (G') is approximately an order of magnitude larger than the loss modulus (G''). This is a typical characteristic of solid-like gel materials. Compared with other peptide or protein hydrogels, Fmoc-FF hydrogel is considerably strong and stiff, and can be stable at a wide range of temperatures and pH values or at extreme acidic conditions. It is thus more advantageous to use Fmoc-FF hydrogel for certain applications such as controlled drug release and three-dimensional cell culture. A molecular model of the Fmoc-FF peptide has been recently proposed to shed light on the molecular stacking mode in the self-assembling fibrous structures. In accordance with this model, the peptides are aligned in an antiparallel β-sheet fashion and adjacent sheets are interlocked through lateral π–π interactions, thus leading to the formation of a cylindrical structure [40].

Low molecular-mass organogelators (LMOGs) are known to gelate a number of organic liquids for the formation of organogels having unique properties and potential applications as distinct soft materials. Recently, we found that the single dipeptide FF can act as a LMOG to gelate some organic solvents such as chloroform and aromatic solvents [41]. The FF organogels are thermoreversible, having an apparent sol–gel transition temperature. SEM (Figure 6.5a and c) and AFM images (Figure 6.5b and d) show that the gel in chloroform is composed of long and flexible fibrils with branches and entangled networks, while except for the prevalent fibrils, there are also ribbon structures in the gel formed in toluene. This implies that different solvents may induce changes of the resulting microstructures in the gel phase or ultimate transition into the crystal phase. Indeed, upon introduction of ethanol as a cosolvent in toluene, the structural transition of organogels into microcrystals is clearly observed [42]. The SEM image (Figure 6.5e) shows that thermodynamic stable structures in the crystal phase are flower-like microcrystals consisting of the packing ribbons. The AFM observation indicates

Figure 6.5 SEM and AFM images (z scale = 100 nm) of the dried FF gels in chloroform (c and e) and toluene (d and f), respectively; SEM image (a) and enlarged AFM three-dimensional height image (b) of the flower-like structure in the microcrystal. (Reproduced with permission from [41]. © 2008, American Chemical Society; also adapted from [42]. © 2010, Wiley.)

that the ribbons have obvious lamellar layers (Figure 6.5d), implying the microcrystals are formed through the hierarchical self-assembly of FF molecules. Intriguing, in 25% ethanol/toluene the dynamic morphological transition from the kinetically trapped state of the gel into the more thermodynamically stable crystal can be directly observed. This provides us with a sound example to help understand the gel–crystal transition process and allows us to obtain structural information in different stages of phase transition.

Multiple spectroscopy techniques, such as Fourier transform IR (FTIR), CD, UV-Vis, and fluorescence spectroscopy as well as XRD and thermogravimetric analysis (TGA) were employed to determine the intermolecular interactions governing the self-assembly of FF in the gels and in the microcrystals. The FTIR spectrum of organogels in toluene showed the predominant β-sheet character based on the position of the amide I band at 1620 and 1683 cm^{-1}, and possibly an antiparallel configuration [43, 44]. The CD spectrum of gels gives a signature with β-sheet arrangements of FF molecules, in agreement with the FTIR analysis. In the crystal phase, FF molecules are possibly more organized in a parallel β-sheet mode based on the amide I band absorption in the vicinity of 1615 cm^{-1} [45, 46]. In the fluorescent spectroscopy measurement, the phenyl groups in the FF solution have an emission peak at 306 nm, which shifts to 339 nm for the gel and 285 nm for the microcrystal [42]. Likewise, the UV absorbance spectrum showed a pronounced obvious shift between gels or microcrystals and free monomers. The red shift in the gel phase suggests an effective π–π stacking between the aromatic residues and probably in a J-aggregate fashion. By contrast, the blue shift in the crystal phase indicates a possible extended H-aggregate between the phenyl rings. The XRD pattern of a dried gel showed a sharp peak at 2θ of 5.2° with a d spacing of 1.7 nm, corresponding to the thickness of the β-sheet monolayer [41]. The microcrystals formed in complete ethanol have a similar XRD pattern to FF single crystals and nanotubes, giving a hexagonal packing structure. Nevertheless, the diffraction patterns of microcrystals obtained in mixed solvents are somewhat different from that in complete ethanol, indicating that toluene molecules possibly interfere or take part in the self-assembly of FF through the aromatic stacking. Such a conjecture is confirmed by TGA of microcrystals in different ethanol/toluene mixed solvents. The weight loss at about 100 °C in the TGA curves is assigned to the toluene molecules embedded in the crystal lattice. The higher the toluene ratio in the mixed solvents, the more weight loss observed at this temperature [42]. On the basis of the results above, it has been suggested that in the gel phase FF molecules may adopt antiparallel β-sheet secondary structures with the J-aggregate nature of aromatic residues and in the crystal state are supposed to organize in the parallel β-sheet form with the creation of H-aggregates in phenyl groups (Figure 6.6). Solvent properties, such as polarity and the ability to form hydrogen bonding, play a vital role in regulating the formation of organogels and controlling the ultimately self-assembled structures, nanofibrils or microcrystals. The discussion of solvent effects on the gelation provides new insights into the solvent–gelator interaction and the molecular arrangement mode in the gel phase as well as in the crystal phase.

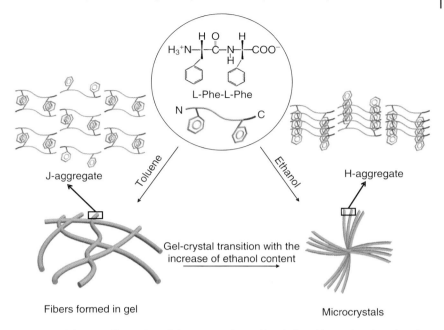

Figure 6.6 Schematic illustration of the structural transition induced by varying the ethanol content in the mixed solvents, and the proposed molecular packing in the gel and in the microcrystal. (Reproduced with permission from [42]. © 2010, Wiley.)

6.2.1.3 Nanowires

FF nanowires vertically aligned on a solid substrate are prepared via solid-phase self-assembly by using an amorphous peptide film as a precursor. Park *et al.* first grew vertically aligned FF nanowires from film by changing the water activity in the vapor phase or by applying high thermal energy [47]. It is proposed that surface nucleation and mass transport limitation may be the main factors for controlling the formation of nanowires on solid surfaces in the water vapor-mediated self-assembly, whereas the nanowires induced by thermal aging are likely a result of a phase transition of FF molecular arrays. Subsequently, they improved the method for the preparation of vertically aligned peptide nanowires from FF amorphous film. With the aid of aniline vapor, uniform and vertically well-aligned nanowires were grown by aging the film at temperatures above 100 °C [48]. SEM images confirmed the occurrence of the vertically well-aligned and rigid nanowires. Based on the analysis of the time evolution of the film through aniline vapor aging, it is suggested that a surface-initiated nucleation in the initial stage is the possible mechanism for the growth of vertically well-aligned nanowires. In addition, a micropattern of vertically aligned FF nanowires can be fabricated by the combination of high-temperature aniline vapor aging and a soft lithographic technique.

FF dipeptide is able to self-assemble into individually dispersed and rigid nanowires in carbon disulfide (CS$_2$) [49]. Such peptide nanowires display novel liquid crystalline behaviors over a wide concentration range. When liquid nanowires are viewed under cross-polarized light, a Schlieren texture indicative of the characteristic morphology of a nematic liquid crystalline phase is observed. Through the exposure to an external electric field, nematic dispersed nanowires can be well aligned. The liquid nanowires from FF-specific self-assembly may find potential for applications in nanopatterning and reinforcing materials into nanocomposites.

6.2.1.4 Ordered Molecular Chains on Solid Surfaces

The fabrication of ordered peptide chains at the molecule level has attracted increasing attention due to the potential applications in bionanotechnology. By using scanning tunneling microscopy (STM), Kern *et al.* directly observed the stereoselective assembly of diphenylalanine enantiomers, L-Phe–L-Phe and D-Phe–D-Phe, into molecule pairs and chains in a chiral recognition fashion [50]. The FF dipeptide contains two chiral carbon centers connected through a central amide bond, which is a key motif in molecular recognition. STM imaging shows that codepositing L-Phe–L-Phe and D-Phe–D-Phe on copper surfaces under vacuum self-organize into homochiral molecule chains through mutually induced conformational changes. The chiral recognition of FF enantiomers at the single-molecule level provides pronounced evidence for the prediction of Pauling on the mechanism of dynamic-induced fit. The two-dimensional extended periodic arrangements of FF molecules can be obtained through the cocrystallization method by using terephthalic acid (TPA) as a linker [51]. STM measurements of FF molecules deposited on copper surfaces reveal that individual FF molecules have a tendency to self-assemble into one-dimensional chains. Through introducing TPA as a molecular "glue" to bridge the isolated dipeptide chains, two-dimensional ordering and extending FF chains emerge on copper surfaces (Figure 6.7). The formation of ordering supramolecular structures is independent of the stoichiometry of the initial FF to TPA, in which molecular organization is self-recognizing with the fixed ratio of FF to TPA (1:1). This makes it easier to produce such ordering nanostructures on solid surfaces owing to not requiring precise control of the amount of deposited components. It has been suggested that intermolecular hydrogen bonding might be the major driving force for the formation of the ultimately ordered structures.

6.2.2
Lipopeptides

A lipopeptide is a molecule consisting of a lipid and a peptide fragment. It has an amphiphilic structure with a hydrophobic tail having a similar structure as lipid hydrophobic segments and a hydrophilic peptide headgroup that incorporates a bioactive or functional sequence. Similar to the self-assembling behavior of naturally occurring phospholipids, lipopeptide molecules, when exposed to an aqueous

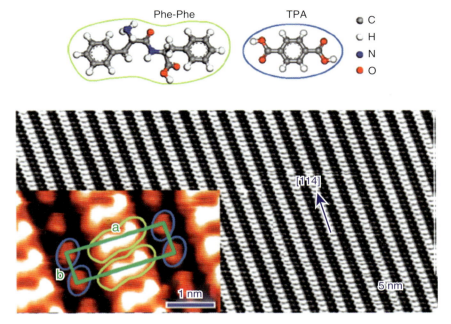

Figure 6.7 STM images of two-dimensional extended FF/TPA ordered molecular chains on Cu(110). The quadrangle in the inset marks one unit of the molecular superlattice. The ellipse and dumbbell point out the TPA and FF molecules, respectively. (Reprinted with permission from [51]. © 2007, American Chemical Society.)

environment, spontaneously aggregate to form vesicular structures resembling liposomes. Within the spherical structure, the lipopeptide molecule is arranged in a bilayer fashion in which the polar hydrophilic peptide headgroups face outwards against the hydrophobic domain from the aqueous environment.

The first pioneering synthesis of lipidated peptides was proposed by Kunitake [52, 53]. The introduction of double- or triple-chain alkyl moieties to one terminus of the peptide sequences produced the amphiphilic molecules. This type of amphiphilic peptide was first used for the purposes of interfacial molecular recognition. Dipeptides modified with N-terminal double-chain hydrophobic moieties could interact selectively with water-soluble dipeptides at the air/water interface. It was proposed that the binding between molecules was a consequence of not only the amenability of host and guests for the formation of hydrogen bonds, but also steric aspects and the placement of the amino acid (hydrophobic) residues. When employing a mixture of amphiphiles with different peptide moieties, a redistribution of monolayer components could be observed upon interaction with guest molecules. The dynamic nature of the interplay between molecules allows for the creation of a binding site suitable for the guest molecule. As this class of peptide amphiphiles consists of a hydrophobic tail, a linker, a spacer, and a

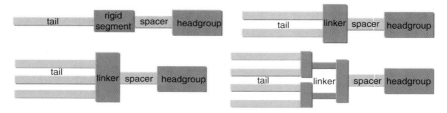

Figure 6.8 Building blocks of peptide amphiphiles. (Reprinted with permission from [52]. © 1992, Wiley.)

hydrophilic head (Figure 6.8), one can control the morphology, characteristics, surface chemistry, and function of the molecules by changing any of the structural segments. Such amphiphiles are flexible for bottom-up fabrication and can self-assemble into a variety of supramolecular architectures (e.g., micelles, vesicles, ribbons, fibers, etc.) [13, 54].

The recent advances in peptide synthesis strategies, particularly the emergence of solid-phase peptide synthesizers, have allowed the preparation of numerous peptide sequences and peptide derivatives at large scale. The solid-phase synthesis approach is powerful in the synthesis of composite molecules, such as the conjugation of hydrophobic lipid chains into peptide sequences [55, 56]. A famous strategy is based on the standard Fmoc protocols for which the first amino acid is immobilized to a resin by coupling the carboxyl, followed by the reaction of the next Fmoc-amino acid to the amine group of the previous one. Using this method, the production of lipopeptides can be easily achieved in a controlled manner. The hydrophobic lipid chain is coupled to peptide sequences at the final last stage. The final lipopeptide can be obtained upon cleavage of product from the resin. Such a synthetic strategy is versatile and flexible for yielding various predesigned lipopeptides. One can control the morphology, characteristics, surface chemistry, and molecule function by varying any of the structural segments of a lipopeptide molecule [57].

Recently, we presented a multivalent cationic lipopeptide (MCL, Figure 6.9a) that was prepared by the solid-phase peptide synthesis method. Such a lipid peptide displays a distinct advantage in gene therapy owing to its biocompatibility, functional expansibility, relative ease of synthesis, and controllable charge density [58]. The peptide head bears two positive charges at physiological pH, which enables a high binding affinity with DNA. Furthermore, the peptide segment is sensitive to enzymes such as trypsin, which allows the escape of DNA from the complex. It has been demonstrated that such a designed lipopeptide can undergo self-assembly to form vesicles in aqueous solution. A confocal laser scanning microscopy (CLSM) image shows that MCL vesicles have a capacity of binding negatively charged DNA. To further verify the attachment of DNA to MCL vesicles, a standard ethidium bromide (EtBr) DNA fluorescence quenching exclusion assay was performed. EtBr is a fluorescent indicator that will fluoresce as it intercalates

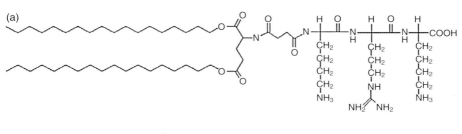

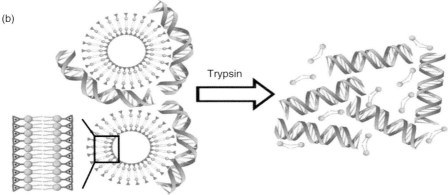

Figure 6.9 (a) Structural formula of the synthesized MCL. (b) Proposed mechanism of DNA release from the MCL/DNA complex after trypsin hydrolysis. (Reprinted with permission from [58]. © 2007, American Chemical Society.)

in the middle of base pairs of DNA, but will self-quench in a free state. However, when DNA binds to MCL vesicles, EtBr can be extruded, which thus can be used to indicate the formation of MCL/DNA complexes. The results showed that the fluorescent intensity of EtBr decreases at the isoelectric region, implying the stable binding of DNA to MCL vesicles. It therefore provides a new way to utilize the cationic lipopeptide vesicle as a carrier for gene delivery and subsequent transfection. Bioactive peptide fragments, such as the short amino acid sequence arginine–glycine–aspartic acid (RGD), having specific recognition capacity for integrin-binding extracellular matrix (ECM) protein, have been successfully incorporated into the headgroup of a lipopeptide via covalent attachment to a lysine residue of a monolysinylated cationic amphiphile for targeted delivery of genes [59]. The RGD motif-modified lipopeptides deliver the genes to cultured cells preferably via the integrin acceptors. The *in vivo* transfection of genes transported by such lipopeptide vectors showed that tumor growth was significantly inhibited through intravenous administration of the electrostatic complex of the RGD-lipopeptide and anticancer gene in mice bearing an aggressive tumor. Tirrell *et al.* prepared biomimetic membranes upon fusion of vesicles functionalized with the lipopeptides containing the RGD motif (Figure 6.10) [60]. Mouse fibroblast cells were observed

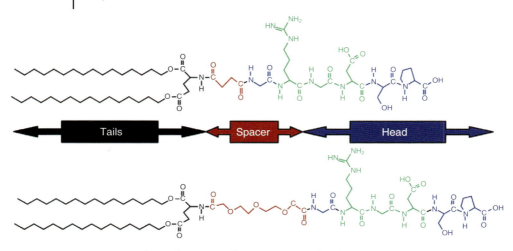

Figure 6.10 Chemical structure of RGD peptide amphiphiles used: $(C_{16})_2$-Glu-C_2-GRGDSP and $(C_{16})_2$-Glu-PEO-GRGDSP. (Reprinted with permission from [60]. © 2007, American Chemical Society.)

to preferably adhere and grow on the surface with modification of RGD-lipopeptides, but dependent on the embedded spacer in the molecule. The poly(ethylene glycol) (PEG) spacer lipopeptides displayed the best capacity for cell adhesion and growth, which was further investigated in a RGD-lipopeptide concentration-dependent manner by creating surface composition arrays using microfluidics. This study provided an effective method to screen biological probes for cell adhesion and growth by setting up peptide composition gradients in a membrane environment.

Peptide sequences can be linked covalently to phospholipids to facilitate the intercalation of lipopeptides into the cell membrane. In biological systems, membrane proteins are anchored to the lipid membrane upon lipidation in co- and posttranslational enzymatic processes such as acylation with fatty acids, prenylation, and rather commonly C-terminal amidation with glycosylphophatidylinositols [61]. Membrane fusion, a key process in all living cells, contributes to the transport of molecules between and within cells. This process is triggered by specific interactions of protein fusion, of which mimicking is important for understanding cell interplay. Nowadays, much research has been performed to setup biomimetic models for the study of membrane fusion. The synthesized phospholipid peptides are indispensable components for preparing model membranes supported on any solid substrates. Marsden *et al.* synthesized two lipidated oligopeptide hybrids (LPE and LPK, Figure 6.11a) serving as a minimum coiled-coil pair that is capable of assembling specifically into a stable heterodimer that possesses all of the functional aspects of membrane-bound SNARE proteins (SNARE = soluble NSF attachment protein receptor; NSF = *N*-ethylmaleimide-sensitive factor)

6.2 Peptides as Building Blocks for the Bottom-up Fabrication of Various Nanostructures

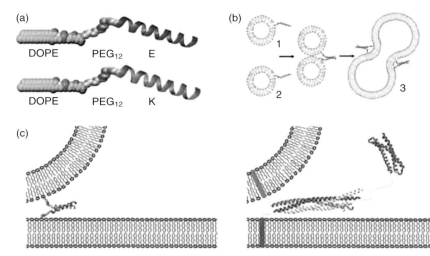

Figure 6.11 (a) Molecular structure of the lipidated oligopeptides LPE and LPK in a space-filling form. Each consists of a DOPE tail linked through a PEG12 spacer to the coiled-coil-forming oligopeptides E and K. The amino acid sequence of E is G(EIAALEK)$_3$-NH$_2$ and that of K is (KIAALKE)$_3$GW-NH$_2$. (b) Spontaneous intercalation of the DOPE tail in lipid bilayers results in liposomes functionalized with either E or K peptides at the surface. As a liposome population bearing LPE (1) is mixed with a liposome population carrying LPK (2), coiled-coil formation (E/K) initiates liposome fusion (3). (c) Comparison of the minimal lipopeptide-based model (left) with the SNARE protein-based model (right). (Reproduced with permission from [62]. © 2009, Wiley.)

(Figure 6.11) [62]. SNARE proteins provide the recognition sites on the lipid bilayers and act by the coiled-coil interaction to drive the fusion of transport vesicles with the neuronal membrane [63]. In the present artificial model, the protein recognition domain was mimicked by two three-heptad repeat coiled-coil-forming peptides (E and K). The formation of the LPE/LPK complex drove two different liposomes close together. A short PEG chain is designed to enable extension of the oligopeptide component from the surface of the liposomes. The lipopeptides are embedded spontaneously in the membrane by the hydrophobic interaction of the phospholipid tail, 1,2-dioleoyl-sn-glycero-3-phosphatidylethanolamine (DOPE) [64], which emulates the function of the transmembrane domain of SNARE proteins. The reduced SNARE model possessing all of the characteristics of native membrane fusion makes this system a minimum for SNARE-mediated membrane fusion. This model system also suggests a potential as a carrier for direct delivery of encapsulated reagents to cells or liposome. The establishment of biomimetic systems based on lipidated peptides attached to a lipid bilayer offers an alternative avenue to gain insights into the most important aspects of membrane fusion [65].

6.2.3
Polypeptides

Polypeptides are a class of polymers formed by linking many molecules of amino acids through amide bonds. Their properties are determined by the type and sequence of the constituent amino acids. Peptide polymers are attractive due to their molecular recognition property, biocompatibility, and wide-ranging application in medicine and biology. Compared to conventional synthetic polymers, polypeptides display many advantages. They can be designed to form different secondary structures in aqueous solution depending on the selection of amino acids. Such a property contributes to the self-assembly of polypeptide chains. However, the preparation of polypeptides (especially long sequences of more than 30 residues) is tedious and expensive. The Deming lab optimized the synthesis of polypeptides via the ring-opening polymerization of α-amino acid–N-carboxyanhydrides (NCAs), which have been available since the 1950s [16, 66–68]. This method was initially suitable for the synthesis of heterogeneous materials, but with poor control over chain length, amino acid composition and sequence, and chain end functionality [67]. With the development of initiators that allow for the living polymerization of NCAs, the fabrication of well-defined materials made from synthetic polypeptides is feasible [16, 68]. Such a polymerization reaction is versatile and flexible, which makes it possible to incorporate various amino acid monomers, both natural and synthetic, resulting in unprecedented flexibility of molecular design.

A general synthetic strategy is shown as in Figure 6.12. Monomer species are first formed by transforming α-amino acid to NCAs, usually in a single step, followed by solution polymerization for which the chain length is controlled by the monomer to initiator stoichiometry. Subsequent addition of different NCA monomers to the active chain ends causes the production of block architectures with defined block lengths and low polydispersity. The use of transition metal initiators in NCA polymerization facilitates the preparation of well-defined homopolypeptides. The synthesis of polypeptides with a narrow molecular weight distribution ($M_w/M_n < 1.20$) and controlled molecular weights ($500 < M_n < 500\,000$) is achieved by using cobalt and nickel as initiators [69]. Orthogonal protection strategies enable precise modifications to amino acid side-chains along the chains, affording conjugates with tailored functions [70]. Based on the above, the ring-opening polymerization of NCAs is an effective strategy for the preparation of a large variety

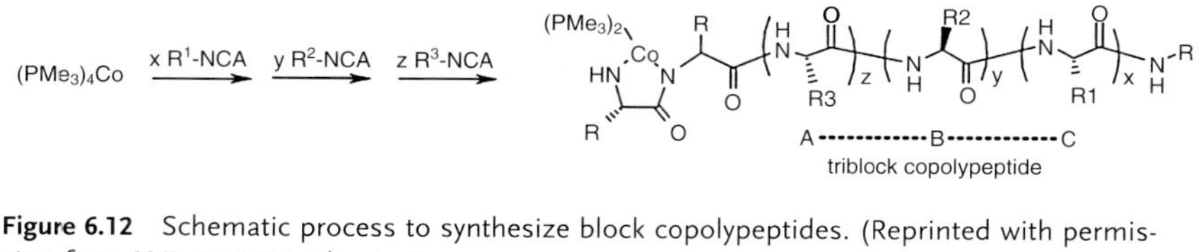

Figure 6.12 Schematic process to synthesize block copolypeptides. (Reprinted with permission from [66]. © 2007, Elsevier.)

of block polypeptides and subsequent application of self-assembled polypeptide superstructures.

Using the NCA polymerization method, Deming *et al.* synthesized a number of copolypeptides with control of chain length and composition for preparing rigid hydrogels in water [17]. Hydrogels are a type of soft material featuring hydrated, porous structures that can mimic natural ECMs [71, 72]. They have great significance for applications in tissue and bone engineering. Artificial polypeptide hydrogels are attractive as candidates for tissue engineering. Deming *et al.* established structure–property relationships in these materials through the association of chemical synthesis and structural characterization, allowing a high level of control over gel porosity, strength, functionality, and media stability [73]. The formation of hydrogels was observed in a series of polypeptides containing a charged, water-dissolving domain and a hydrophobic domain having defined secondary structures. The gelation process depends not only on the amphiphilic nature of polypeptides, but also on the type of secondary structures present in chain. It was found that α-helices or β-strands contribute to the formation of hydrogels, while random coils inhibit the gelation. The highly charged water-dissolving segments were designed as cationic poly(L-lysine) (PLL)-HBr and poly(L-glutamic acid) (PGA), sodium salt, while hydrophobic domains were designed to adopt a regular conformation, as α-helices for poly(L-leucine) and β-sheets for poly(L-valine). The formation of hydrogels is relevant to the molecular parameters and composition of polypeptides. K160L40 is a typical polypeptide used to form stable hydrogels by adopting a α-helix conformation in hydrophobic segments. By increasing the amount of leucine in the polypeptide, the gel strength can be enhanced dramatically; however, owing to the increasing electrostatic repulsion, longer hydrophilic segments can lead to distortion of hydrophobic helices, in essence preferring to form flat two-dimensional sheets [74]. A model is presented in Figure 6.13 to describe the helix packing of

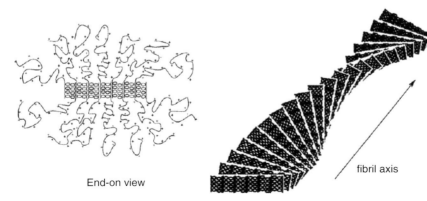

Figure 6.13 Drawings showing the proposed packing of block copolypeptide amphiphiles into twisted fibrillar tapes, with helices packed perpendicular to the fibril axes. PLL chains were omitted from the fibril drawing for clarity. (Reprinted with permission from [66]. © 2007, Elsevier.)

polypeptides for minimizing the system energy. In this model, the helices are stacked perpendicular to the fibril axis, but with a slight twist between planes of parallel packed helices. Through TEM observations, supramolecular structures in K380L20 are more fibrillar, tape-like nanostructures constituting the hydrogel network. Replacement of cationic PLL segments in K160L40 with anionic PGA segments (E160L40) has no effect on hydrogel formation [17].

A polypeptide multilayer film can be fabricated on the surface of many solid substrates by the layer-by-layer (LbL) assembly technique, which is a general approach for the preparation of multilayers. As the name implies, the polypeptide multilayer film is made of peptide polymers [75]. In some cases, another type of polymer is involved in the fabrication process for either adjusting physical properties or adding new functions, such as a chemically modified polypeptide [76], a nonbiological organic polyelectrolyte [77], or a polysaccharide [78]. The preparation of polypeptide multilayer films is basically consistent with the process for fabricating polyelectrolyte multilayers in which oppositely charged species are assembled on a solid substrate via alternate deposition (Figure 6.14). Actually, polypeptides for LbL assembly are a type of weak polyelectrolyte, the linear charge density of which is adjustable by simply varying the pH. The pK_a of ionizable groups in a polypeptide is sensitive to the local electronic environment and the net charge may shift significantly from the solution value on formation of a polyelectrolyte complex or film [79–81]. We can see that polypeptide multilayer films are not only biological benign, but also function-adjustable.

Multiple characterization techniques are available for studying LbL multilayers. SEM and AFM are suitable for studying the surface morphology of multilayer films;

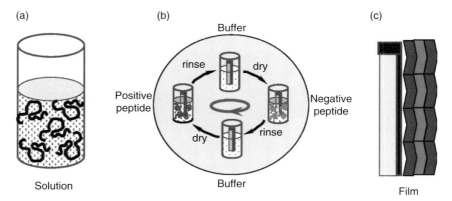

Figure 6.14 Schematic illustration of polypeptide multilayer film fabrication by the LbL assembly technique. Oppositely charged polypeptides in solution (a) are adsorbed alternately onto a solid support (b), such as a quartz slide, yielding a multilayer film (c). Loosely bound components are rinsed off in water or buffer solution in the process of assembly. The film is in principle required to dry after each adsorption step for the next adsorption, which also allows some measurement. (Reproduced with permission from [75]. © 2005, American Chemical Society.)

in particular, AFM is a more generally used tool to probe surface roughness, thickness, and topological structures not only for dried films, but also wet ones [82–88]. CLSM has commonly been applied to supervise the fluorescent film – to observe the diffusion of polypeptides within multilayers [78], to study film biodegradation [89], and to visualize hollow polypeptide microcapsules [90]. FTIR spectroscopy [87, 90, 91] and CD spectrometry [83, 84, 92] are widely used to obtain information on the secondary structure of polypeptide multilayer films. In particular, CD enables a moderately accurate determination of film secondary structure content. The far-UV CD signal is very sensitive to the conformation of the polypeptide backbone. By contrast, CD is a more accurate means to determine peptide secondary structure because it can give more distinctive spectral signatures than FTIR. UV-Vis spectroscopy is a relatively simple and cost-accessible tools to measure the optical mass of polypeptides in accordance with the absorbance increase per layer [83, 84]. FTIR spectroscopy can also be used to measure the optical mass of assembled materials by detecting chemical bond vibrations in an attenuated total reflection mode [93–96]. Ellipsometry [83, 97, 98] and optical waveguide light-mode spectroscopy [82, 83, 92–95] are applicable for the measurement of optical film thickness and refractive index. The amount of polypeptides adsorbed can be calculated from the thickness and refractive index. Quartz crystal microbalance (QCM), an acoustic technique, is an eminent means for obtaining information on the mass change and kinetics of polyelectrolyte adsorption by monitoring the change of resonant frequency [83, 84, 92, 99]. Viscoelastic properties of hydrated films can be studied by dissipative QCM. Electrical properties of a film surface can be studied by a streaming potential method, the basic principle of which is to measure the pressure and potential difference on both sides of a capillary.

Polypeptide multilayer films are promising for the development of applications owing to their biocompatibility, biodegradability, specific biomolecular sensitivity, environmental advantages, antifouling characteristics, thermal responsiveness, and stickiness or nonstickiness [75]. Composition of the polypeptide multilayers is flexible, which is easily controlled over the synthesis of polypeptides with a flexible change of amino acid sequences. Therefore, corresponding variations of multilayer functions can result. Due to their biological nature, polypeptide LbL films show many possible applications in medicine and bionanotechnology, such as enantiomeric separations [100, 101], antimicrobial films [102], artificial skin grafts [103], cell and tissue culture [82, 104], immunogenicity control [90], implant technology [97], and artificial cells and drug delivery systems [105, 106].

Polypeptide nanotubes can be constructed by the LbL assembly technique in combination with a porous template [107]. These biological nanotubes have some inherent advantages for the application in drug or gene delivery. Polypeptide secondary structure and biological function can be readily controlled by varying the peptide sequence in the process of synthesis, which thus leads to versatile functional materials [75, 108]. To achieve the fabrication of controllable polypeptide nanotubes, PLL-HCl and PGA are selected as the model molecules. It is known that PLL is a polycation, while PGA is a polyanion. Similar to the assembly process of polyelectrolytes in porous membrane templates, polypeptide multilayers can be

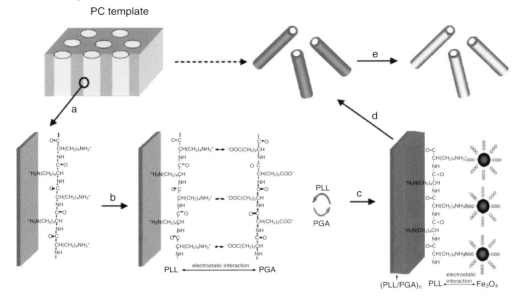

Figure 6.15 Illustration of the preparation of magnetic polypeptide nanotubes by using the LbL assembly technique combined with a porous template. (a) Adsorption of a single polypeptide layer; (b) assembly of polypeptide multilayers on a polycarbonate membrane template; (c) adsorption of Fe_3O_4 nanoparticle/PLL bilayers; (d) dissolution of the PC membrane that results in liberation of the membrane-supported nanotubes; (e) adsorption of negatively charged DNA on the outer surface of as-assembled nanotubes. (Reprinted with permission from [107]. © 2008, Royal Society of Chemistry.)

formed by alternately adsorbing PLL and PGA due to electrostatic interaction (Figure 6.15). Compared with the self-assembly method, the LbL assembly technique has the advantage that the wall thickness of nanotubes can be adjustable and thus can control their mechanical properties. Such nanotubes can also be endowed with new functions upon the introduction of functional components. For instance, magnetic polypeptide nanotubes are constructed by depositing magnetite nanoparticles into the inner surface of PLL/PGA multilayers on the template (Figure 6.15c). More interestingly, such functional nanotubes have an ability to bind negatively charged drugs or genes (Figure 6.15e) and also have an intrinsic biodegradability owing to the fact that peptide materials can be hydrolyzed by a number of enzymes. Therefore, there is the potential to use such functional well-defined nanotubes as carriers for site-specific drug or gene delivery.

6.2.4
Amphiphilic Peptides

Amphiphilic peptides are peptide molecules comprised of a hydrophobic domain (amino acids only or an alkyl tail) and a hydrophilic peptide segment. Such a peptide has a typical characteristic that it can spontaneously aggregate in water to form assemblies such as micelles. Amphiphilic peptides are very attractive as building blocks for bottom-up fabrication, and have potential application in biotechnology and materials science.

Inspiration from nature is helpful to direct the design of peptide building blocks towards self-assembly. For instance, based on the discovery of a repetitive 16-residue peptide motif n-AEAEAKAKAEAEAKAK-c (EAK16-II), a fragment of a left-handed Z-DNA-binding protein in yeast, Zhang et al. designed a class of ionic self-complementary peptides such as RDA16-I, RAD16-II, and EAK-I [5]. This class of peptide building block shares a common architectural feature that both positively and negatively charged side-chains are on one side of the β-sheet and hydrophobic side chains on the other. They have been shown to undergo self-assembly into nanofiber scaffolds, having potential application in three-dimensional tissue cultures. Inspired by the self-assembly of lipids, a class of essential components in cell membranes, Zhang et al. designed surfactant-like oligopeptides with hydrophobic tails and hydrophilic heads, including negatively charged Ac-AAAAAAD (A_6D), Ac-VVVVVVD (V_6D), Ac-VVVVVVDD (V_6D_2), and Ac-LLLLLLDD (L_6D_2), and positively charged Ac-AAAAAAK-NH_2 (A_6K), Ac-VVVVVVK-NH_2 (V_6K), Ac-VVVVVVH-NH_2 (V_6H), Ac-VVVVVVKK-NH_2 (V_6K_2), Ac-LLLLLLKK-NH_2 (L_6K_2), and so on. (Figure 6.16) [110]. These amphiphilic peptides have adjustable hydrophobic tails with different degrees of hydrophobicity composed of certain hydrophobic amino acids such as glycine, alanine, valine, leucine, or isoleucine. Furthermore, the hydrophilic heads can be designed to be either negatively or

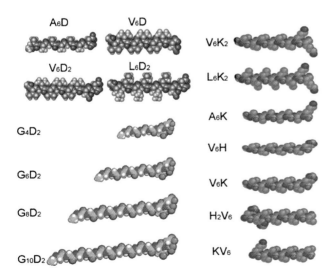

Figure 6.16 Molecular structures of designed surfactant-like peptides. D bears negative charges, K and H bear positive charges, and A, V, and L constitute the hydrophobic tails with increasing hydrophobicity. Each peptide is about 2–3 nm in length, similar to biological phospholipids. (Reprinted with permission from [109]. © 2009, Elsevier.)

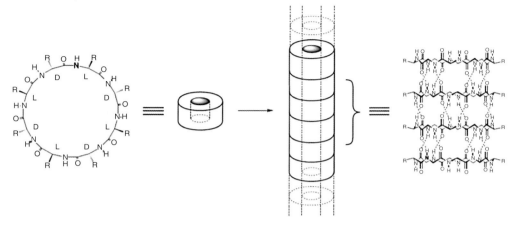

Figure 6.17 Schematic diagram of nanotube assembly from cyclic D,L-peptides. For clarity, most of the peptide side-chains have been omitted. (Reproduced with permission from [10]. © 2001, Wiley.)

positively charged depending on the selection of charged acids. They can undergo self-assembly by the aggregation of hydrophobic tails in water similar to a lipid molecule, but a possible difference for driving self-assembly still exists compared to natural lipids because the surfactant-like peptides also likely interact through the hydrogen bonds along peptide backbones [110].

In 1993, Ghadiri *et al.* designed a cyclic D,L-peptide motif arranged with alternating D- and L-amino acids for the creation of a hollow tubular structure by the stacking of cyclic peptide rings [111]. The sequence of octapeptide cyclo[-(L-Gln–D-Ala–L-Glu–D-Ala)$_2$-] was selected to render solubility in basic aqueous solution and thus to prevent subunit association through Coulombic repulsion. Controlled acidification allows for production of microcrystalline aggregates with high aspect ratios. The formation of tubes is demonstrated owing to the ordered stacking of ring-shaped subunits via antiparallel β-sheet hydrogen bonds (Figure 6.17). After that, a variety of cyclic peptides made from D,L-α-amino acids, D,L-β-amino acids, alternating α- and β-amino acids, alternating α- and γ-amino acids, δ-amino acids, enantiopure cyclo-*N*,*N*'-linked oligoureas, and so on, have been designed for the fabrication of tubular structures [112–118]. Such cyclic peptide building blocks are readily synthesized and subjected to multiple rounds of sequence–activity optimizations through either parallel or combinatorial library approaches. In the cyclic form, pore size is well defined and controllable, which thus contributes greatly to the incorporation of function. Functionality can be added to the self-assembling supramolecular scaffolds by placing functional moieties into the side-chains of cyclic peptides at molecular precision. Such nanotubular structures have many possible applications ranging from the preparation of novel cytotoxic and controlled-release drug delivery agents to catalytic and materials science applications, such as biomineralization and site isolation of chromophores or other reactive groups [10].

The α-helix has proven to be a helpful building block for bottom-up fabrication [119]. For instance, the leucine zipper (LZ) motif, a specific type of dimeric α-helical coiled-coil, is well-known for the fabrication of nanostructured biomaterials. LZs are small motifs (less than 30 amino acids in length) accessible synthetically; when folded, LZs have the approximate dimensions of 4 nm × 2 nm, cylinders and finally scalable LZs as tandem LZ repeats can be used to self-assemble into nano- or micrometer regimes, but still reflect the underlying nanoscale tectons [120]. Woolfson *et al.* reported that a redesigned LZ peptide with 28 residues could be manipulated into fibers, known as the first-generation self-assembling fiber system [121]. In this system, two complementary LZ peptides are designed to coassemble to form sticky-ended dimmers that, in turn, are engineered to foster end-to-end assembly into long, noncovalently linked α-helical coiled-coil fibrils. Subsequently, they engineered a second-generation self-assembling fiber based on α-helical coiled-coil building blocks associated with a heptad repeat (**abcdefg**), where hydrophobic residues at the **a** and **d** sites can facilitate the formation of a hydrophobic core of the coiled-coil helical bundle. The helical interface was further connected with interhelical charge–charge interactions between sequential **g** and **e** sites. Specifically, discrete, complementary charge clusters were introduced to the surfaces of the interacting coiled coils, namely a pair of negatively charged aspartates on one peptide, to supplement an identically spaced pair of positively charged arginine side chains on the other, eventually achieving the sticky-ended heterodimer programming for the self-assembly of fibers [122]. It is noteworthy that the coiled-coil motif [123] is an α left-handed superhelix, consisting of two or more right-handed α-helices, which exists extensively in proteins ranging from muscle proteins to DNA transcription factors. It has typical amino acid heptad repeating units designated as **a** to **g** in one helix and **a′** to **g′** in the other. The hydrophobic residues locate at the interface of the two helices (**a**, **d** and **a′**, **d′**), while **e**, **g** and **e′**, **g′**, exposed to solvent, are polar residues that lead to the specificity between the two helices through electrostatic interactions (Figure 6.18).

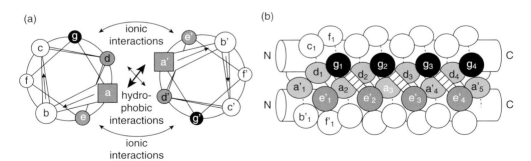

Figure 6.18 Schematic of a parallel dimeric coiled-coil. The helical wheel diagram in (a) is a top view down the axis of the α-helices from the N- to C-terminus; (b) is a side view. The residues are labeled **a–g** in one helix and **a′–g′** in the other. (Reproduced with permission from [123]. © 2004, Wiley.)

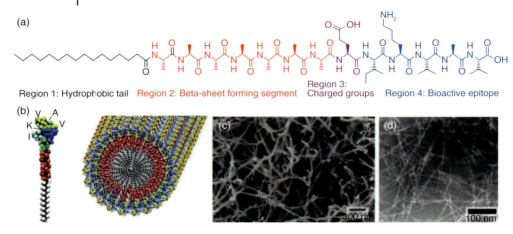

Region 1: Hydrophobic tail Region 2: Beta-sheet forming segment Region 3: Charged groups Region 4: Bioactive epitope

Figure 6.19 (a) Chemical structure of a representative peptide amphiphile with a rational design for the fabrication of biomaterials. (b) Schematic illustration of an IKVAV-containing peptide amphiphile and its self-assembly into nanofibers. (c) SEM and (d) TEM images of the IKVAV nanofiber network formed by adding cell media (Dulbecco's modified Eagle's medium) to the peptide amphiphile aqueous solution. (Reproduced with permission from [128]. © 2010, Wiley.)

Stupp *et al.* designed and developed a class of peptide amphiphiles, consisting of a hydrophobic alkyl tail covalently coupled to a hydrophilic peptide sequence, for construction of self-assembling biomaterials [12, 124–127]. These molecules show the structural features of amphiphilic surfactants functionalized with bioactive peptides. The amphiphilic nature allows for outward specific presentation of hydrophilic peptide signals when self-assembled into defined nanostructures in aqueous solution. A typical chemical structure for these peptide amphiphile molecules is shown in Figure 6.19(a). They contain three key structural segments: (i) a hydrophobic moiety, which is commonly an alkyl tail; (ii) a short peptide sequence promoting the formation of nanofibers through intermolecular hydrogen bonding; (iii) sometimes charged amino acids were introduced to enhance solubility in water and for the design of pH- or salt-responsive assemblies; and (iv) a peptide segment containing ionizable residues, usually a bioactive motif of interest for biological signaling. These peptide amphiphile molecules can pack in an ordered fashion to form energy-favorable long fibers (Figure 6.19b–d). Hydrophobic interactions of the alkyl tails, hydrogen bonding among the middle peptide segments, and electrostatic repulsions between the charged amino acids are responsible for driving the formation of high-aspect-ratio fibrous structures [128]. The first two are attractive forces that contribute to the aggregation of peptide amphiphiles, whereas the electrostatic repulsion from the charged components isolates the peptide amphiphile molecules, which possibly make them grow orientationally along a one-dimensional direction. The size, shape, and interfacial curvature of the final assemblies likely reflect a precise balance of these forces.

The ultimate peptide amphiphile nanostructures and their gelation behavior can be tuned via the control of molecular forces, which can be implemented through design of molecular structures, adjustment of self-assembly conditions, and introduction of other coassembling molecules. For example, Hartgerink et al. [127] investigated the effect of amino acid sequence and the length of alkyl tail on nanofiber formation. Results indicated that peptide amphiphile molecules with long alkyl tails, such as C_{10}, C_{16}, and C_{22}, formed self-supporting hydrogels under slow acidification, whereas peptide amphiphile molecules with shorter hydrophobic tails of C_6 or with no tail did not result in nanofibers at any pH. At pH 7, peptide amphiphile molecules had a net negative surface charge that was gradually offset with the addition of acids, which thus allows for the self-assembly of nanofibers. The formation and structure of nanofibers was independent of the starting concentration of peptide amphiphiles, but the bundle of fibers could be observed at higher concentrations. The formation of bundles was regarded as necessary for the formation of self-supporting gels. Aside from the induction of acids, it was found that a divalent cation, such as Ca^{2+}, could also initiate the gelation of peptide amphiphiles. Subsequently, they studied the role of hydrogen bonding and amphiphilic packing in the self-assembly process of nanofibers [129]. It was demonstrated that it was necessary for the four amino acids closest to the hydrophobic alkyl tail to be linked through β-sheet hydrogen bonds to drive the formation of nanofibers. Instead of elongated fibrous structures, spherical micelles were formed when these hydrogen bonds were disrupted.

Stupp et al. also showed that oppositely charged peptide amphiphiles with different bioactive sequences coassembled into nanofibers through electrostatic attraction between the two kinds of peptide amphiphiles [130, 131]. The introduction of a diacetylene group into hydrophobic segments of peptide amphiphile molecules can stabilize the fiber architecture through polymerization following the self-assembly. It was found that the polymerization yielded highly conjugated backbones when the peptide moiety of peptide amphiphiles had a linear, opposed to a branched, structure [132]. The polymerization of monomers in peptide amphiphiles provided indirect information on the ordered organization of the internal core. Peptide amphiphile nanofibers with a rigid conjugated backbone had considerable mechanical robustness and insolubility important for their patterning at a cellular scale. The internal structure or organization in peptide amphiphiles was efficiently studied by fluorescence techniques [133–135]. Probe molecules such as tryptophan or pyrene chromophore were conjugated to the peptide backbone and the end of the alkyl tail of peptide amphiphile molecules, which allowed for the spectroscopic examination of the interior of the resulting supramolecular objects. These studies showed that the interior of nanofibers remained well solvated by water, but more densely packed and geometrically confined near the hydrophobic core compared to the periphery. More recently, they achieved control of the mechanical properties of peptide amphiphile nanofiber gels by varying the amino acid sequences that commonly tend to form β-sheet hydrogen bonds [136]. By manipulating the number and position of valines and alanines in the peptide sequence, they could tailor the stiffness of the gel (where valines contributed to increase the gel rigidity, while additional alanines weakened the mechanical properties). Vitreous ice cryo-TEM, FTIR, and CD spectroscopy were employed to investigate the morphology of self-assembling fibers, and the relationship between internal structures of peptide amphiphile fibers with different amino acid sequences and gel stiffness. Insight into the molecular factors affecting gel stiffness is important in controlling these matrices for specific biological targets.

6.3
Peptide–Inorganic Hybrids

A combination of inorganic functional materials (such as nanocrystals, nanoparticles, and polyoxometalates (POMs)), which exhibit unique electronic, photonic, and catalytic properties, with biological or biomimetic building blocks provides a strategy towards improved properties and novel functions of nanobiomaterials. Remarkable efforts have been to identify suitable organic components to achieve integration with functional inorganic materials, particularly nano-objects. Peptides, as versatile self-assembled building blocks with unique biological functionalities, such as recognition properties, possess remarkable potential for the fabrication of such multifunctional hybrids. They could be applied to direct the growth and assembly of inorganic nanostructures. Significant progress has been made in the association of peptide building blocks with functional inorganic components either for the organization of inorganic nanostructures based on structured peptide templates or for the construction of functional hybrid materials [137].

6.3.1
Nonspecific Attachment of Inorganic Nanoparticles on Peptide-Based Scaffolds

Nowadays, the assembly of inorganic nanoparticles through peptide-based approaches has attracted great attention due to the development of peptides as building blocks for bottom-up fabrication. In particular, their inherent biological originality and important roles played in biomineralization make them attractive in the design and creation of a variety of inorganic nanoparticle superstructures, including one-dimensional nanoparticle chains [138–140], two-dimensional nanoparticle sheets [141], three-dimensional nanoparticle spheres [142–144], and nanoparticle double helices [145]. Use of self-assembling peptide scaffolds for the assembly of nanoparticle superstructures involves the following independent steps: (i) peptide assembly, (ii) synthesis of nanoparticles, (iii) nanoparticle binding with peptide-based scaffolds, and finally (iv) formation of the nanoparticle assemblies templating the peptide scaffolds. In some cases, one can control the direct nucleation and growth of nanoparticles on the peptide scaffolds – known as *in situ* biomineralization.

It is well known that electrostatic repulsion is commonly indispensible for stabilizing nanoparticles in aqueous solutions. Such charged nanoparticles may be used as building blocks to assemble superstructures by combining oppositely charged peptide-based supramolecular architectures. In this case, electrostatic interaction (or attraction) becomes a driving force to tune the interplay of particles and peptide-based templates. For instance, Wang et al. [138] designed and fabricated three-dimensional peptide–inorganic nanoparticle hybrid scaffolds by using positively charged peptide nanofibers, which were assembled from T1 peptide RGYFWAGDYNYF beforehand, as a template in association with negatively charged metal nanoparticles through electrostatic interaction. They conjectured that the protonation of T1 peptide played a key role in directing the assembly of negatively gold particles. The resulting hybrid superstructures were dependent of nanoparticle sizes. It was found that small gold nanoparticles (3.6 nm) tended to form double-helical arrays, whereas large particles, in particular those whose average diameter was larger that 10 nm, randomly attached to peptide nanofibers. The double helices were formed at pH 6, under the conditions at which a T1 peptide was speculated to assemble in a helical structure form. The block copolypeptides, poly(EG$_2$-K)$_{100}$-b-poly(D)$_{30}$, could be used to direct the assembly of 6-nm magnetic nanoparticles (γ-Fe$_2$O$_3$) into water-soluble clusters [142]. The studies revealed that electrostatic interactions between the positively charged nanoparticles and negatively charged carboxylic acid residues of the poly(Asp)$_{30}$ played a favorable role to facilitate the assembly of nanoparticles. Similarly, the group of Stucky and Deming fabricated robust and hollow spheres composed of two distinct layers of silica and gold nanoparticles with the aid of the lysine–cysteine diblock copolypeptides [143].

By using a similar strategy, the group of Pochan and Schneider constructed two-dimensional sheets of gold nanoparticles adopting self-assembled nontwisted laminated β-sheet peptide fibrils as temples [141]. *De novo* designed peptide, (VK)$_4$-VPPT-(KV)$_4$, was used to assemble laminated β-sheet fibrils consisting of alternating layers of hydrophobic valine residues and hydrophilic lysine residues (Figure 6.20). Such a unique peptide nanostructure is suitable for use as a nanoscale template. Negatively charged gold nanoparticles (2.0 ± 0.3 nm in diameter) were mixed with preformed peptide fibrils, resulting in periodically spaced, parallel, linear nanoparticle arrays (Figure 6.20b). The ordered organization of nanoparticles on the two-dimensional sheets was predominately ascribed to the electrostatic interaction between negatively charged particles with the positively charged lysine residues, but was affected by multiple factors such as the nontwisted laminated morphology of a template, strong complementary interaction, and the same order between particle size and lateral spacing of the fibril [141]. Recently, the group of Pochan and Kiick fabricated one-dimensional gold nanoparticle chains with precise axial separations by using self-assembling polypeptide fibrils that have a unique molecular architecture with spatially repeating positively charged patches [139]. An alanine-rich polypeptide was precisely designed for the formation of fibrils with regularly spaced charged patches along the longitudinal axis of the fibril. Such charged patches have an ability to bind oppositely charged inorganic nanoparticles, which thus result in linear nanoparticle arrays. This strategy is able to be extended to construct one-dimensional arrays of different nanoparticle types

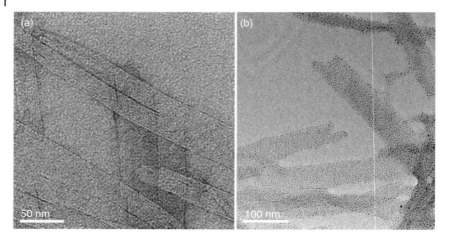

Figure 6.20 TEM images of laminated β-sheet fibrils stained with uranyl acetate (a) and peptide fibril–gold nanoparticle hybrids (b). (Reproduced with permission from [141]. © 2008, Wiley.)

and internanoparticle spacing can be modulated by programming the polypeptide sequence [139].

The hybrid two- or three-dimensional architectures with the incorporation of functional inorganic nanocrystals are of fundamental interest for application in nanotechnology. Such spatial network scaffolds can be readily created by using a gelator to gelate the corresponding well-dispersed solutions containing nanocrystals. It has been found that FF is a suitable candidate to achieve the fabrication of three-dimensional fibrous scaffolds [41]. Upon simply gelating the solution of quantum dots (QDs), organogels with entrapped QDs are easily obtained. Such gels display photoluminescent properties, which is a contribution of the QDs themselves (Figure 6.21a). TEM images show that gels with encapsulated QDs consist of cross-linking three-dimensional fibrous networks with attachment of QDs on the fibrils (Figure 6.21b and c). The comparison of the emission spectra of QDs in gels and those of free QDs reveals that the emission maxima of the QDs in fibrous networks are slightly blue-shifted, indicating the attachment of QDs on fibrils, but the original photoluminescent colors remain (Figure 6.21d). In addition, lipophilic gold nanoparticles are manipulated into the fibrous scaffolds by using the same method, indicating the general suitability of this approach. It can thus be predicted that gel materials with various optical, electronic, and magnetic properties may be achieved through gelating the corresponding functional nanocrystals.

Recently, a new strategy was initiated to prepare water-dispersable three-dimensional colloidal spheres by using FF-based organogels with encapsulated nanocrystals as a starting point [144]. This method is defined as a transfer of a nanocrystal organogel phase into a water phase (TNOW). The detailed processes

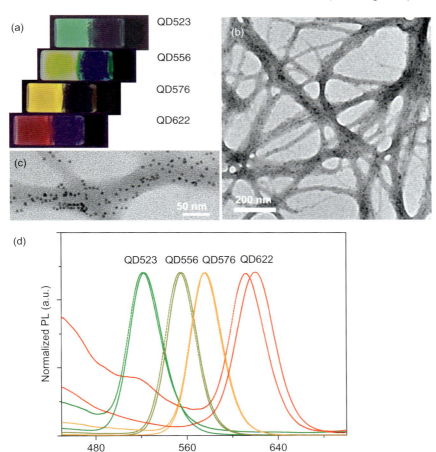

Figure 6.21 Incorporation of QDs in gel. (a) Photoluminescence photograph of four different encapsulated QDs gels. (b) TEM image of the encapsulated QD523 nanocrystals in the fibril network. (c) Magnified TEM image showing the attachment of QDs to the fibril. (d) Emission spectra ($\lambda_{excitation}$ = 365 nm) of the free QDs in toluene (solid line) and the encapsulated QDs in gel (dash dot). (Reprinted with permission from [41]. © 2008, American Chemical Society.)

are illustrated in Figure 6.22(a). The FF derivative, cationic dipeptide (H-Phe–Phe-NH$_2$·HCl), serves as the gelator to initially prepare the fibrous aggregates of QDs (Figure 6.22b) that are then dried to get the xerogel with the attached QDs under vacuum. By the addition of water and subsequent ultrasonic treatment, the resultant three-dimensional colloidal spheres can be obtained. TEM images shows that the colloidal spheres comprise the individual QDs (Figure 6.22c), which ensures the original optical properties of the QDs after assembly into three-dimensional

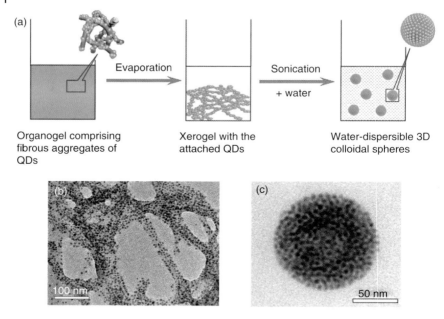

Figure 6.22 (a) Schematic illustration for the preparation of water-dispersible three-dimensional colloidal spheres. TEM images of the fibrous aggregates of QDs (b) and an as-prepared colloidal sphere (c). (Reproduced with permission from [144]. © 2008, Wiley.)

colloidal spheres. The local energy-dispersive X-ray (EDX) spectrum further confirms that the colloidal spheres are composed of QDs and cationic dipeptide. Dynamic light scattering (DLS) study indicates that the colloidal spheres are stable in aqueous solution, having hydrodynamic diameters of about 150 ± 100 nm. Furthermore, we can also obtain different sized colloidal spheres relying on the concentration of cationic dipeptide and ultrasonic time. It has been proposed that the electrostatic interactions upon the protonation of the cationic dipeptide and hydrophobic attraction are possibly the predominant driving forces for the assembled process.

6.3.2
Peptide-Based Biomineralization

Biomineralization is the process by which living organisms can produce minerals, commonly to harden existing tissues. It is ubiquitous in biological system. The studies and understanding of hard tissue growth and regeneration are quite extensive, which on the one hand help us gain a deeper insight into biological processes for the production of minerals, and on the other hand inspire us to design and develop biomimetic materials under mild and environmentally benign conditions. With inspiration from natural biomineralization process, many scientists have contributed great efforts to the creation of functional hybrid materials by using biological molecules, including proteins [146, 147] and peptides [137, 147]. In particular, peptides have attracted more attention owing to their versatility for bottom-up fabrication, flexibility for introduction of biological or chemical function, and ease of synthesis by genetic engineering or chemical routes.

Development of molecular biology methods now allows for the combinatorial selection of peptides with high affinities for specific inorganic species. Through such methods, a number of specific inorganic-binding peptides have now been identified for the fabrication of a variety of inorganic nanostructured materials. These peptide sequences are specific for directing the nucleation and growth of inorganic nanostructures. Aside from naturally occurring peptides as suitable candidates for biomineralization, some synthetic peptides obtained by combinatorial library approaches (e.g., phage display and cell-surface display) enrich the diversity of specific inorganic-binding peptides that have significant advantages for the preparation of unnatural inorganic materials [3]. For example, the peptide sequence SLKMPHWPHLLP or TGHQSPGAYAAH was found to specifically bind germanium for the preparation of amorphous germania nanoparticles at room temperature [148]. By using phage-display methods in conjugation with polymerase chain reaction, Naik *et al.* discovered a cobalt-binding peptide Col-P10 (HYPTL-PLGSSTY) that promoted the production of discrete CoPt nanoparticles with an average diameter of 3.5 ± 0.5 nm [149].

With a step-by-step understanding of the mechanism of biomineralization, especially the relationships between peptide sequences and their binding specificities and affinities, scientists have designed and synthesized many novel peptides capable of specifically binding inorganic species. For instance, due to the fact that nucleation and growth of $CaCO_3$ crystals are relevant for the appearance of carboxylate-rich proteins, Sewald *et al.* designed a class of amphiphilic peptides consisting of alternating hydrophilic (Asp) and hydrophobic (Phe) amino acid residues to mimic the epitopes of acidic proteins for biomineralization of $CaCO_3$ [150]. In addition, a variety of mineralizing peptides were bound to the sidewalls of peptide nanotubes assembled from bola-amphiphilic peptides (e.g., bis(*N*-α-amidoglycylglycine)-1,7-heptane dicarboxylate) for direct synthesis of nanocrystals on the nanotubes [22]. In this case, the coating peptide sequences served as the nucleation sites to specifically synthesize inorganic nanoparticles templating the tube morphology. For example, the histidine-rich peptide AHHAHHAAD (HRE) was immobilized on nanotubes for the growth of gold nanocrystals in the presence of a reducing agent [151]. Likewise, copper nanocrystals were anchored to the nanotubes with incorporation of peptide HGGGHGHGGGHG (HG12), and diameters of the resulting nanocrystals could be varied by controlling HG12 conformation and aqueous pH [152]. The HG12 peptide was also applicable for the growth of Cu_2S semiconductor nanocrystals on amphiphilic peptide nanotubes [152]. Shape control of silver nanocrystals on nanotubes was achieved by using NPSSLFRYLPSD (AG4) as the mineralizing peptide, for which AG4 recognized and affected the silver nanocrystal growth kinetics on the (111) face, thus leading to an isotropic coating of hexagonal silver nanocrystals on the nanotubes [153].

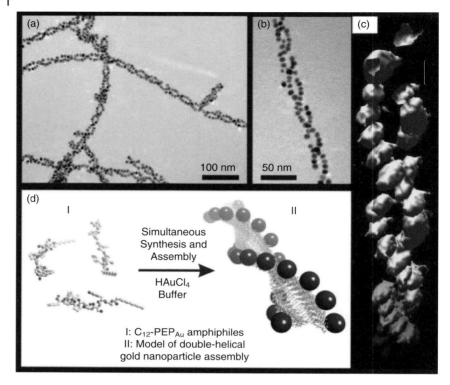

Figure 6.23 (a and b) TEM images of left-handed gold nanoparticle double helices. (c) Tomographic three-dimensional reconstruction image of left-handed gold nanoparticle double helices. (d) Schematic illustration of the formation of left-handed double helices. (Reproduced with permission from [137]. © 2010, Wiley.)

More recently, Rosi et al. [145] used an amphiphilic peptide termed C_{12}-PEP_{AU} (C_{12}-AYSSGAPPMPPF) to achieve the simultaneous peptide self-assembly and biomineralization of gold nanoparticles. Intriguingly, they obtained highly ordered left-handed double helices of gold nanoparticles with a length even up to the microscale level (Figure 6.23).

Hydroxyapatite (HA, $Ca_5(PO_4)_3(OH)$), a naturally occurring mineral form of calcium apatite, can be found extensively in both teeth and bone. Actually, HA and its derivatives can be prepared artificially by using various methods such as precipitation reactions [154] and sol–gel synthesis [155], but a significant challenge remains to direct the growth of HA capable of mimicking biological systems. Self-assembling supramolecular systems offer the possibility to be basic models for mimicking the architecture of fibrous matrices. A well-controlled process of biomineralization in synthetic systems, which emulates protein-mediated mineralization, can be realized by using precisely designed peptide molecules having self-assembling properties. Stupp et al. have done a lot of work on biomimetic mineralization of bone apatite using peptide amphiphile building blocks. Such peptide supramolecular structures allow not only for incorporation of bioactive epitopes, but also for chemistry specifically targeting mineralization processes [156]. A bone biomimetic mineralization process was achieved by peptide amphiphile nanofibers formed by self-assembly. As mentioned in Section 6.2.4, peptide amphiphiles are a class of amphiphilic peptides consisting of a hydrophobic alkyl tail linked to a functional peptide sequence, commonly bearing charged amino acid residues. The peptide moiety of the peptide amphiphile molecule aiming at bone mineralization was modified using a phosphoserine that could interact strongly with calcium ions to direct mineralization of HA, while an RGD sequence was also introduced as a cell adhesion ligand [12]. Self-assembling peptide nanofiber networks were formed by controlling aqueous pH and further stabilized by cross-linking of disulfide bonds through oxidation. Mineralization of HA crystals was directed by templating the fiber scaffolds, forming a hybrid material having a similar organization observed between collagen fibrils and HA in bone. The above mineralization of HA was performed in a two-dimensional porous surface containing self-assembling nanofibers. In a following study, they extended this biomimetic process to a three-dimensional gel matrix that could be used for bone regeneration (Figure 6.24) [157]. A phosphorylated anionic peptide amphiphile nanofiber gel matrix was used to template, in three dimensions, HA nanocrystals with size, shape, and crystallographic orientation resembling the

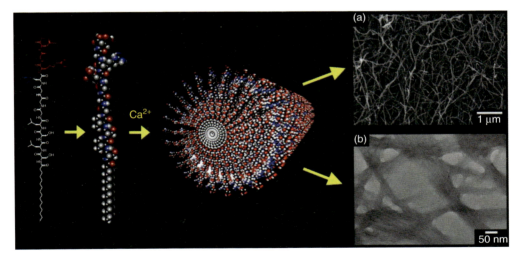

Figure 6.24 Schematic illustration of RGD-peptide amphiphile and their self-assembly into a nanofiber. Low-magnification (a) and high-magnification (b) SEM images and the TEM image (c) show fibrous bundles, made up of peptide amphiphile nanofibers approximately 5–7 nm in diameter. The SEM images were taken of a critical-point dried peptide amphiphile gel, while the TEM image was taken of nanofibers dried on a TEM grid and stained with phosphotungstic acid. (Reproduced with permission from [157]. © 2009, Wiley.)

natural bone mineral. The HA mineralization was enzymatically tuned and thus promoted by harvesting of phosphate ions. It was suggested that both temporal and spatial templating are necessary to yield the biomimetic nanocrystals. Gradually, enzymatic release of phosphate ions regulated the availability of precursors, which thus affected the rate of crystal nucleation. Such a controlled production of mineral precursors prevented uncontrolled mineral precipitation in bulk. The further studies suggested that the peptide nanofibers played a key role in directing the observed crystallographic alignment and templated mineralization of HA in the three-dimensional system. This strategy provides a new avenue for the creation of biomimetic materials to promote bone regeneration.

6.3.3
Adaptive Hybrid Supramolecular Networks

Aside from nano-objects, POMs, a known class of anionic oxide nanoclusters of transition metals, are suitable candidates for the fabrication of peptide-based functional hybrids owing to their potential application in catalysis, electronics, optics, magnetics, medicine, and biology [158]. A Keggin-type POM (phosphotungstic acid (PTA)) was selected as a polyoxoanion model molecule and combined with the cationic dipeptide to assemble the expected hybrids [159]. The morphology of the hybrids was systematically investigated by using SEM, TEM, and DLS techniques. The results show well-defined spherical nanostructures with diameters of around 150 ± 60 nm (Figure 6.25b). EDX spectroscopy in combination with SEM and FTIR indicated that the hybrids consisted of PTA and cationic dipeptide. Based on the XRD and high-resolution TEM characterization, it has been suggested that the hybrid spheres are formed upon initial self-assembly of peptide encapsulated clusters (PECs) by strong electrostatic interaction and further stacking of such

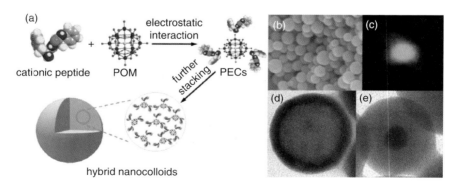

Figure 6.25 (a) Schematic illustration of coassembly of cationic dipeptide and POM into hybrid spheres. (b) SEM image of hybrid spheres. Incorporation of guests into hybrid spheres: (c) CLSM image of FITC-doped colloids, (d) TEM image of colloids encapsulated with Hypocrellin B nanoparticles, and (e) TEM image of colloids with the inclusion of gold nanoparticles. (Reprinted with permission from [159]. © 2010, Wiley.)

PECs through multiple noncovalent interactions (Figure 6.25a). Such hybrid spheres are responsive to external stimuli such as pH and temperature, which is a desirable feature for self-assembled nanostructures for extensive application in the controlled release of drugs.

Intriguingly, the hybrid supramolecular network displays an adaptive inclusion property for guests in the self-assembly process, which is utilized to encapsulate various guest materials [159]. Water-soluble small molecules including neutral fluorescein isothiocyanate (FITC), positively charged rhodamine 6G, and negatively charged Congo Red as well as macromolecular polymers such as dextran labeled with FITC (FITC–dextran) have been demonstrated to enable the effective incorporation into the hybrid spheres. For example, in a CLSM measurement, the hybrid spheres with encapsulation of FITC can emit clear fluorescence (Figure 6.25c). Such a hybrid supramolecular network also represents the interfacial adaptability for nanoscale guests, regardless of the hydrophobic or hydrophilic interface. The hybrid supramolecular network is demonstrated to self-assemble adaptively on the surface of hydrophobic drug particles and form core–shell nanostructures with the drug particle as the core (Figure 6.25d). Additionally, hydrophilic gold nanoparticles are incorporated into the hybrid supramolecular network and result in Au@hybrid core–shell spheres (Figure 6.25e). The hybrid supramolecular network self-assembled from the association of POM and cationic dipeptide exhibits flexible properties and multifunctional behavior, offering a new insight into hybrid supramolecular structures, and may be a starting point for further seeking and fabricating bioinorganic hybrid materials with novel functions.

6.4
Applications of Peptide Biomimetic Nanomaterials

The production of various functional nanostructures based on peptide building blocks provides a broad prospect for the potential applications from biological to nonbiological fields. The biological applications involve tissue engineering, delivery of drugs or genes, bioimaging and biosensors, and so on. In addition, self-assembled peptide nanostructures can serve as excellent templates for the fabrication of functional inorganic or organic materials. As a typical example, FF nanotubes, nanowires, and ribbons have been used to prepare metal nanowires, functional polymer nanotubes, and hollow TiO_2 ribbons. Such peptide-based templates are susceptible to enzymes and physical or chemical treatment, which enables them to be readily removed from the resultant materials.

6.4.1
Biological Applications

6.4.1.1 Three-Dimensional Cell Culture Scaffolds for Tissue Engineering
There is an increasing interest in culturing cells in three dimensions, imitating a more "natural" environment (i.e., the ECM of structural proteins and other

biological molecules found in living tissues) [160]. To provide a three-dimensional environment suitable for growing cells, it is necessary to design and fabricate three-dimensional fibrous scaffolds that are able to promote the adhesion and differentiation of cells. A number of peptide-based fibrous networks have been developed to mimic an artificial ECM. For example, Zhang *et al.* have designed a series of peptide building blocks with alternating hydrophobic and hydrophilic residues (where the hydrophilic residues, in turn, alternate between being positively and negatively charged, such as in $(KLDL)_n$, $(EAKA)_n$, and $(RADA)_n$) for the construction three-dimensional hydrogel networks [161–164]. Interestingly, these designer peptides could self-assemble into fibrous networks in physiological media, which was a primary condition to grow cells. The synthesized fibrous peptide scaffolds have nanoscale pores and an extremely high water content (above 99.5% in weight), which closely emulate the porosity and gross structure of the ECM. It thus provided an appropriate accommodation for cells in three-dimensional environments, and allowed the growth factors and nutrients to diffuse in and out very slowly. These types of peptides have been shown to be noncytotoxic and have potential in the repair of cartilage tissue. When chondrocytes were seeded in three-dimensional hydrogel scaffolds, the fibrous scaffolds maintained the differentiated chondrocytes, and stimulated the synthesis and accumulation of the ECM. For the first time they achieved the fabrication of *de novo*-designed scaffolds and with potential use for three-dimensional cell culture.

Stupp *et al.* at Northwestern University developed a series of peptide amphiphiles containing a hydrophobic alkyl tail for tissue regeneration purposes, including cartilage repair and promotion of nerve cell growth. The hydrophilic peptide head was at the molecular scale functionalized with laminin epitope IKVAV [165]. (Laminin is an ECM protein that affects neurite outgrowth.) Such peptide amphiphile molecules functionalized with a bioactive motif were manipulated to assemble into fibrous hydrogel networks for encapsulation of neural progenitor cells. In comparison with individual laminin and soluble peptides, the three-dimensional hydrogel scaffold induced very rapid differentiation of cells into neurons, while suppressing the development of astrocytes. Peptide amphiphile three-dimensional scaffolds have also been used to serve as a model to study the effects of matrix geometry on the behavior of cells. Stupp *et al.* adapted microfabrication and lithography techniques to the fabrication of well-defined micro- and nanoscale patterns of peptide amphiphile nanofibers [166–168]. Peptide amphiphile molecules containing both a photopolymerizable portion and the RGDS cell-adhesive peptide were allowed to self-assemble within microfabricated molds. Such confined, self-assembly allowed for the creation of either randomly oriented or aligned nanofiber networks with well-defined topographical patterns such as holes, posts, or channels up to 8 mm in height and down to 5 mm in lateral dimensions [166]. These patterned surfaces were used to culture human mesenchymal stem cells (MSCs), monitoring the effects of topographical and nanoscale features on the migration, alignment, and differentiation of these cells. It was found that osteoblastic differentiation was enhanced on hole microtextures in the topographical patterns with randomly oriented nanofibers, while the topographical patterns containing peptide amphiphile nanofibers aligned through flow prior to gelation promoted the alignment of MSCs. These patterned structures fabricated by lithography combined with self-assembly may have potential for application in biomedical devices. In another example, MC3T3-E1 cells were grown in a peptide amphiphile nanofibrous matrix that was triggered to self-assembly by polyvalent metal ions under physiological conditions [169]. The studies revealed that MC3T3-E1 cells entrapped in peptide fibrous matrices could survive for at least 3 weeks, and that entrapment did not arrest cell proliferation and motility. Such peptide amphiphile molecules amenable to self-assembly under physiological conditions have significant potential for use in cell transplantation and tissue engineering.

FF-based nanomaterials have also shown potential in three-dimensional cell culture. For example, fibrous hydrogel networks have tentatively been used as the ECM for the three-dimensional culture of cells. Chondrocyte cells were used as model cells and incorporated into the hydrogel by mixing with appropriate Fmoc-FF gel. Two-photon fluorescence microscopy and environmental SEM studies reveal the features of a number of chondrocytes, which indicate the growth and proliferation of cells in the three-dimensional fibrous networks [38]. Therefore, FF-based fibrous scaffolds are promising for further development for application in tissue engineering, which enables them to ultimately function analogous to the natural ECM. The Ulijn group [170] developed a FF-based hydrogel with the incorporation of a biorecognition function by coassembling the Fmoc-FF and Fmoc-RGD peptides. Such hydrogels may be used as biomimetic fibrous scaffolds for the three-dimensional culture of anchorage-dependent cells.

6.4.1.2 Delivery of Drugs or Genes

As stated in Section 6.2.2, the lipopeptide vesicles can bind and compact DNA, and are thus expected to be effective gene transfection reagents for application in gene therapy. For this purpose, plasmid DNA is mixed with MCL and the resulting complexes are incubated with HeLa cells for 16 h. Flow cytometry is used to determine the transfection efficiency. In the presence of MCL, the transfection efficiency is about $45 \pm 5\%$, which is twice that of commercial transfection reagents [58]. Furthermore, cytotoxicity assays showed that MCL has no obvious influence on cell growth. The mechanism for gene delivery and release in cells is proposed in Figure 6.9(b). First, the plasmid DNA binds to the headgroups of the cationic peptide of MCL vesicles. Then, the MCL/DNA vesicles enter the HeLa cells (e.g., via endocytosis) and trypsin in the cells can hydrolyze the peptide of the plasmid DNA/MCL complexes. The charge reversal of the vesicles will repel DNA and assist DNA to escape from the endosome for subsequent transcription [171, 172]. This process is readily realized and thus enhances DNA transfection efficiency. Hence, such a synthesized MCL offers a new alternative vector for the delivery of therapeutic DNA.

Polypeptide vesicles assembled from (poly(L-arginine)$_{60}$-b-poly(L-leucine)$_{20}$, $R_{60}L_{20}$) diblock copolypeptides have been developed as a vehicle for drug delivery [173]. These vesicles were able to entrap water-soluble molecules or drugs, such as, Texas-Red-labeled dextran, and they could be internalized to both epithelial (T84) and endothelial (HULEC-5A) cell lines, and were found to be minimally cytotoxic. The guanidinium residues of arginines are essential components for the effective intracellular delivery of vesicles and cargoes. This is because the arginine-rich sequences mimic the function of the protein-transduction domains (PTDs), which are able to enhance the intracellular delivery of cargoes [174–176]. Recently, Tirrell *et al.* studied the interactions of micelles composed of a palmitoylated, proapoptotic peptide derived from p53 tumor suppressor protein with cells *in vitro* [177]. Although elongated rod-like micelles could be formed above a critical micelle concentration, monomers instead of aggregated micelles were actually internalized to cells. This was ascribed to the nature of dynamic self-assembly of such peptide micelles and the weak noncovalent interactions that hold them together. The internalization of peptide amphiphiles was shown to occur through adsorptive-mediated energy-dependent endocytosis pathways. This study suggested an efficient way to increase peptide permeability inside cancer cells (SJSA-1 human osteosarcoma cell line). Improving stability of micelles, such as by using polymerizable peptide amphiphiles, is required for intact micelle internalization.

It has been demonstrated that FF-based nanomaterials have various potential applications in the biological field. FF-based nanostructures show potential in the delivery of drugs or genes *in vitro*. A recent example showed that oligonucleotides can be delivered to the interior of cells by endocytosis interactions upon the spontaneous transition of self-assembled CDPNTs into vesicles by varying the concentration of building blocks (Figure 6.26) [30]. CLSM and gel electrophoresis studies confirm the secure immobilization of negatively charged oligonucleotides on the CDPNTs and subsequent intracellular internalization of reassembled vesicles. Furthermore, such peptide-based delivery vehicles are biocompatible, bioabsorbable, and recyclable. It thus is possible to exploit the FF-based nanomaterials as a new class of vectors for the delivery of many foreign substances.

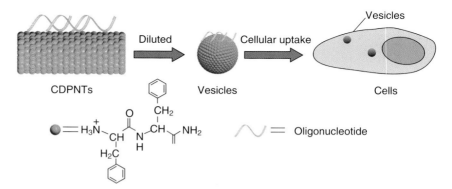

Figure 6.26 Schematic illustration of oligonucleotide delivery into cells by using the peptide carrier. (Reprinted with permission from [30]. © 2007, Wiley.)

6.4.1.3 Bioimaging

Bioimaging function of FF-based nanomaterials can be achieved through the integration of functional inorganic nano-objects into the peptide system. Three-dimensional colloidal spheres upon the association of cationic dipeptide and QDs exhibit the capability of bioimaging (Figure 6.27) [144]. As illustrated in Section 2.3, such three-dimensional colloidal spheres are fabricated by a new strategy called the TNOW method. They can be stably dispersed in a serum-containing cell medium. A standard cytotoxicity assay, known as the 3-(4,5-dimethylthiazolyl-2)-2,5-diphenyltetrazolium bromide cell survival assay, shows that the three-dimensional colloidal spheres are highly biocompatible, offering evidence for further application in cell labeling. CLSM studies indicates that after incubation with three-dimensional colloidal spheres the cells can radiate punctate spots of

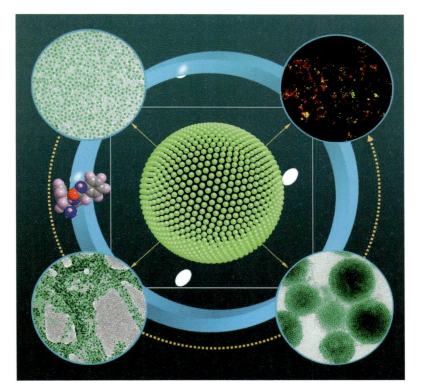

Figure 6.27 Bioimaging application of water-dispersible three-dimensional colloidal spheres prepared by the combination of cationic dipeptide and QDs. Lipophilic QDs (top left) can be transformed into QD-containing fibrils (bottom left) via addition of cationic dipeptide. By adding water and subsequent sonication these can be converted into three-dimensional colloidal spheres (bottom right), which then can be internalized to cells for bioimaging (top right). Note that top right is a fluorescence micrograph, the others are TEM images.

luminescence. A red FM 4-64 marker used to stain cell membranes and endosomes showed the location of three-dimensional colloidal spheres inside cells. The result indicates that three-dimensional colloidal spheres are internalized into cells and possibly accumulated in the cytoplasm. Therefore, this investigation presents an excellent example to attain functionality of biomaterials upon their integration with functional inorganic materials, which leads to the new functions and thus enlarges their applications.

6.4.1.4 Biosensors

Peptide nanotubes self-assembled from FF building block have been shown to have applications in biosensors [178, 179]. A new electrochemical biosensing platform was fabricated by depositing FNTs on the surface of a graphite electrode [178]. Cyclic voltammetric and time-based amperometric studies demonstrate that the presence of FNTs can significantly improve the sensitivity of electrodes. In a similar study, thiol-modified FNTs were applied to a gold electrode, and made it sensitive to enzymes such as glucose oxidase and ethanol dehydrogenase [179]. Such modified electrodes show improved sensitivity and reproducibility for the detection of glucose and ethanol based on the enzyme-related electrocatalytic reaction. In addition, FNT-modified electrodes also have some other merits such as nonmediated electron transfer, short detection time, large current density, and comparatively high stability. Therefore, these findings show that FNTs are an attractive alternative for the fabrication of sensors and biosensors having promising electrochemical performance.

6.4.2 Nonbiological Applications

Peptide nanostructures have an exciting potential application in nanofabrication, in which nanoscale nanotubes, nanowires, or ribbons serve as templates for the formation of functional nanomaterials such as metal nanowires and polymer (e.g., polyaniline (PANI)) nanotubes, and so on. As a know example, water-filled FNTs have been deemed as a favorable scaffold for creating metal nanowire and composites embedded with metal nanoparticles (Figure 6.28). Gazit *et al.* have confirmed that FNTs can be used as a template for metallization [21]. A silver nanowire 20 nm in diameter was obtained upon reduction of silver ions inside the nanotubes followed by enzymatic degradation of peptide scaffolds. Through the introduction of thiol-containing peptide linkers, 20-nm gold nanoparticles were able to be coated on the surface of the FNTs filled with silver nanowires, resulting in the attainment of metal–peptide–metal trilayer coaxial nanocables with unique electromagnetic properties [180]. In a similar method, a nanotube fabricated from the FF dipeptide was used to template platinum nanoparticles [33]. High-resolution TEM and EDX spectrum studies demonstrated the attachment of 2–3 nm platinum nanoparticles in the walls of peptide nanotubes and the following formation of platinum nanoparticle–peptide nanotube composites.

The FF xerogel composed of highly entangled ribbons can serve as a biotemplate to enable the construction of highly complicated hierarchical architectures via a

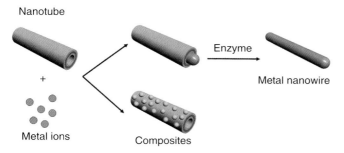

Figure 6.28 Illustration of FNTs as functional templates for the fabrication of metal nanowires and composites embedded with metal nanoparticles.

mild, biologically benign process [181]. Kim *et al.* fabricated a hollow TiO$_2$ ribbon through templating the FF ribbon framework, which was first self-assembled in chloroform from the FF peptide [182]. Such peptide ribbons have rather high thermal stability, which can thus allow them to undergo a functional process of atomic layer deposition (ALD) at high temperature (140–160 °C). A thin TiO$_2$ layer was spread over the peptide ribbon scaffolds by using such an ALD method. Since the peptide template can be removed by calcination, highly entangled hollow TiO$_2$ ribbons that replicate the FF organogel morphology may readily be produced. In contrast to the majority of biomolecularly assembled structures, the ribbons constructed from the aromatic dipeptide of FF exhibit remarkable potential for nanofabrication under much more rigorous conditions. Vertically aligned FF nanowires on a solid substrate have shown an ability to act as a template for the fabrication of FF/PANI core–shell conducting nanowires [183]. The thickness of the PANI shell can be readily adjusted by either varying the reaction time or the number of PANI layers coated. Electrochemical analyses demonstrate that the FF/PANI nanowires are active both chemically and electrochemically. Individual PANI nanotubes can also be obtained by selectively removing the FF template. This investigation indicates that FF-based assemblies are versatile and practical for the fabrication of functional polymer nanostructures.

Superhydrophobic surfaces comprised of FF nanowires were fabricated by the self-assembly of FF with exposure to fluorinated aniline vapor at high temperature [184]. It is proposed that both nanoscale roughness of the peptide nanowire film and decrease of surface free energy are responsible for the superhydrophobic properties. The self-cleaning superhydrophobic surface fabricated upon peptide self-assembly opens up a new possible application for peptide-based materials.

6.5
Conclusions and Perspectives

Molecular assembly such as self-assembly and the LbL assembly method, as a powerful "bottom-up" technique, has been broadly applied in the fabrication of biological and biomimetic materials. Peptides and proteins as a class of assembly-enabling biomolecules exhibit significant potential for the construction of such biological nanomaterials. Generally, one can discover and synthesize some peptide building blocks with inspiration from natural motifs and their self-assembling behaviors. A variety of supramolecular architectures, including nanotubes, spherical vesicles or liposome, nanofibers, and ribbons, can be constructed by self-assembly of rationally designed peptide building blocks. In addition, the integration between such peptide molecules and functional inorganic components is readily achieved, which thus improves the inherent properties of biomaterials and adds new functions. A number of applications are available for such peptide-based nanomaterials. For instance, fibrous hydrogel scaffolds constructed from various self-assembling peptides can serve as the ECM for cell culture in a three-dimensional environment. Also, biocompatible three-dimensional colloidal spheres with incorporation of QDs can be used to label living cells, and peptide–POM hybrid spheres have an adaptive incorporation property and thus can be used to encapsulate a wide range of guest materials. FF nanotubes and ribbons can serve as a biological templates for creating functional inorganic materials such as metal nanowires, metal oxide ribbons, and so on. Lipopeptides and polypeptides have been shown to be able to manipulate liposome-like vesicles and functional multilayer nanotubes by the molecular assembly method, respectively. In particular, lipopeptide vesicles exhibit considerably potential as an effective vector for the delivery of therapeutic DNA. It is thus believed that such biomimetic materials will show significant potential as a class of innovative tool for the novel delivery of drugs or genes.

Although significant advances have been made in the creation of peptide-based biological and biomimetic materials, this field is still in its infancy. So far, there seems to be a notable lack of theoretical insight into peptide nanostructure formation, such as rational design of peptide motifs for self-assembly, and effective prediction and precise control of the properties of the resulting structures. Fabrication of materials through the incorporation of the multiple responsive elements or components is still a challenge. Thus, it still remains a formidable challenge for the future development and application of peptide materials. As the future research focus of peptide building blocks, one may envisage that the introduction of functional units into peptide entities will offer an effective way for improving the intrinsic properties and tuning the self-assembly, and ultimately achieving multifunctional nanobiomaterials for practical application in biology and nanotechnology.

References

1 Lehn, J.M. (2002) Toward self-organization and complex matter. *Science*, **295**, 2400–2403.
2 Sanchez, C., Arribart, H., Madeleine, M., and Guille, G. (2005) Biomimetism and bioinspiration as tools for the design of innovative materials and systems. *Nat. Mater.*, **4**, 277–288.
3 Sarikaya, M., Tamerler, C., Jen, A.K., Schulten, K., and Baneyx, F. (2003)

Molecular biomimetics: nanotechnology through biology. *Nat. Mater.*, **2**, 577–585.

4 Aizenberg, J. and Fratzl, P. (2009) Biological and biomimetic materials. *Adv. Mater.*, **21**, 387–388.

5 Zhang, S.G. (2003) Fabrication of novel biomaterials through molecular self-assembly. *Nat. Biotechnol.*, **21**, 1171–1178.

6 Seeman, N.C. and Belcher, A.M. (2002) Emulating biology: building nanostructures from the bottom up. *Proc. Natl. Acad. Sci. USA*, **99**, 6451–6455.

7 Tu, R.S. and Tirrell, M. (2004) Bottom-up design of biomimetic assemblies. *Adv. Drug Deliv. Rev.*, **56**, 1537–1563.

8 Venkatesh, S., Byrne, M.E., Peppas, N.A., and Hilt, J.Z. (2005) Applications of biomimetic systems in drug delivery. *Expert Opin. Drug Deliv.*, **2**, 1085–1096.

9 Gazit, E. (2007) Self-assembled peptide nanostructures: the design of molecular building blocks and their technological utilization. *Chem. Soc. Rev.*, **36**, 1263–1269.

10 Bong, D.T., Clark, T.D., Granja, J.R., and Ghadiri, M.R. (2001) Self-assembling organic nanotubes. *Angew. Chem. Int. Ed.*, **40**, 988–1011.

11 Percec, V., Dulcey, A.E., Balagurusamy, V.S.K., Miura, Y., Smidrkal, J., Peterca, M., Nummelin, S., Edlund, U., Hudson, S.D., and Heiney, P.A. (2004) Self-assembly of amphiphilic dendritic dipeptides into helical pores. *Nature*, **430**, 764–768.

12 Hartgerink, J.D., Beniash, E., and Stupp, S.I. (2001) Self-assembly and mineralization of peptide–amphiphile nanofibers. *Science*, **294**, 1684–1688.

13 Kokkoli, E., Mardilovich, A., Wedekind, A., Rexeisen, E.L., Garg, A., and Craig, J.A. (2006) Self-assembly and applications of biomimetic and bioactive peptide amphiphiles. *Soft Matter*, **2**, 1015–1024.

14 Vauthey, S., Santoso, S., Gong, H., Watson, N., and Zhang, S. (2002) Molecular self-assembly of surfactant-like peptides to form nanotubes and nanovesicles. *Proc. Natl. Acad. Sci. USA*, **99**, 5355–5360.

15 von Maltzahn, G., Vauthey, S., Santoso, S., and Zhang, S. (2003) Positively charged surfactant-like peptides self-assemble into nanostructures. *Langmuir*, **19**, 4332–4337.

16 Deming, T.J. (1997) Facile synthesis of block copolypeptides of defined architecture. *Nature*, **390**, 386–389.

17 Nowak, A.P., Breedveld, V., Pakstis, L., Ozbas, B., Pine, D.J., Pochan, D., and Deming, T.J. (2002) Rapidly recovering hydrogel scaffolds from self-assembling diblock copolypeptide amphiphiles. *Nature*, **417**, 424–428.

18 Ulijn, R.V. and Smith, A.M. (2008) Designing peptide based nanomaterials. *Chem. Soc. Rev.*, **37**, 664–675.

19 Hamley, I.W. (2007) Peptide fibrillization. *Angew. Chem. Int. Ed.*, **46**, 8128–8147.

20 Rauk, A. (2009) The chemistry of Alzheimer's disease. *Chem. Soc. Rev.*, **38**, 2698–2715.

21 Reches, M. and Gazit, E. (2003) Casting metal nanowires within discrete self-assembled peptide nanotubes. *Science*, **300**, 625–627.

22 Gao, X.Y. and Matsui, H. (2005) Peptide-based nanotubes and their applications in bionanotechnology. *Adv. Mater.*, **17**, 2037–2050.

23 Reches, M. and Gazit, E. (2004) Formation of closed-cage nanostructures by self-assembly of aromatic dipeptides. *Nano Lett.*, **4**, 581–585.

24 Görbitz, C.H. (2006) The structure of nanotubes formed by diphenylalanine – the core recognition motif of Alzheimer's β-amyloid polypeptide. *Chem. Commun.*, 2332–2334.

25 Kol, N., Adler-Abramovich, L., Barlam, D., Shneck, R.Z., Gazit, E., and Rousso, I. (2005) Self-assembled peptide nanotubes are uniquely rigid bioinspired supramolecular structures. *Nano Lett.*, **5**, 1343–1346.

26 Niu, L.J., Chen, X.Y., Allen, S., and Tendler, S.J.B. (2007) Using the bending beam model to estimate the elasticity of diphenylalanine nanotubes. *Langmuir*, **23**, 7443–7446.

27 Hendler, N., Sidelman, N., Reches, M., Gazit, E., Rosenberg, Y., and Richter, S.

28 Reches, M. and Gazit, E. (2006) Controlled patterning of aligned self-assembled peptide nanotubes. *Nat. Nanotechnol.*, **1**, 195–200.

29 Hill, R.J.A., Sedman, V.L., Allen, S., Williams, P.M., Paoli, M., Adler-Abramovich, L., Gazit, E., Eaves, L., and Tendler, S.J.B. (2007) Alignment of aromatic peptide tubes in strong magnetic fields. *Adv. Mater.*, **19**, 4474–4479.

30 Yan, X.H., He, Q., Wang, K.W., Duan, L., Cui, Y., and Li, J.B. (2007) Transition of cationic dipeptide nanotubes into vesicles and oligonucleotide delivery. *Angew. Chem. Int. Ed.*, **46**, 2431–2434.

31 Yan, X.H., Cui, Y., He, Q., Wang, K.W., Li, J.B., Mu, W.H., Wang, B.L., and Ou-yang, Z.C. (2008) Reversible transitions between peptide nanotubes and vesicle-like structures including theoretical modeling studies. *Chem. Eur. J.*, **14**, 5974–5980.

32 Bhargava, P., Tu, Y.F., Zheng, J.X., Xiong, H.M., Quirk, R.P., and Cheng, S.Z.D. (2007) Temperature-induced reversible morphological changes of polystyrene-block-poly(ethylene oxide) micelles in solution. *J. Am. Chem. Soc.*, **129**, 1113–1121.

33 Song, Y.J., Challa, S.R., Medforth, C.J., Qiu, Y., Watt, R.K., Pena, D., Miller, J.E., van Swol, F., and Shelnutt, J.A. (2004) Synthesis of peptide-nanotube platinum-nanoparticle composites. *Chem. Commun.*, 1044–1045.

34 Li, M. and Ou-yang, Z.C. (2008) Concentration-induced shape transition of nano-aggregates in solution. *Thin Solid Films*, **517**, 1424–1427.

35 Naito, H., Okuda, M., and Ou-yang, Z.C. (1993) Equilibrium shapes of smectic-A phase grown from isotropic phase. *Phys. Rev. Lett.*, **70**, 2912–2915.

36 Hardy, J. and Selkoe, D.J. (2002) The amyloid hypothesis of Alzheimer's disease: progress and problems on the road to therapeutics. *Science*, **297**, 353–356.

37 Cherny, I. and Gazit, E. (2008) Amyloids: not only pathological agents but also ordered nanomaterials. *Angew. Chem. Int. Ed.*, **47**, 4062–4069.

38 Jayawarna, V., Ali, M., Jowitt, T.A., Miller, A.E., Saiani, A., Gough, J.E., and Ulijn, R.V. (2006) Nanostructured hydrogels for three-dimensional cell culture through self-assembly of fluorenylmethoxycarbonyl-dipeptides. *Adv. Mater.*, **18**, 611–614.

39 Mahler, A., Reches, M., Rechter, M., Cohen, S., and Gazit, E. (2006) Rigid, self-assembled hydrogel composed of a modified aromatic dipeptide. *Adv. Mater.*, **18**, 1365–1370.

40 Smith, A.M., Williams, R.J., Tang, C., Coppo, P., Collins, R.F., Turner, M.L., Saiani, A., and Ulijn, R.V. (2008) Fmoc-diphenylalanine self assembles to a hydrogel via a novel architecture based on π–π interlocked β-sheets. *Adv. Mater.*, **20**, 37–41.

41 Yan, X.H., Cui, Y., He, Q., Wang, K.W., and Li, J.B. (2008) Organogels based on self-assembly of diphenylalanine peptide and their application to immobilize quantum dots. *Chem. Mater.*, **20**, 1522–1526.

42 Zhu, P.L., Yan, X.H., Su, Y., Yang, Y., and Li, J.B. (2010) Solvent induced structural transition of self-assembled dipeptide: from organogels to microcrystals. *Chem. Eur. J.*, **16**, 3176–3183.

43 Elfrink, K., Ollesch, J., Söhr, J., Willbold, D., Riesner, D., and Gerwert, K. (2008) Structural changes of membrane-anchored native PrP^C. *Proc. Natl. Acad. Sci. USA*, **105**, 10815–10819.

44 Lamm, M.S., Rajagopal, K., Schneider, J.P., and Pochan, D.J. (2005) Laminated morphology of nontwisting β-sheet fibrils constructed via peptide self-assembly. *J. Am. Chem. Soc.*, **127**, 16692–16700.

45 Barth, A. and Zscherp, C. (2002) What vibrations tell us about proteins. *Q. Rev. Biophys.*, **35**, 369–430.

46 de Groot, N.S., Parella, T., and Aviles, F.X. (2007) Ile–Phe dipeptide self-assembly: clues to amyloid formation. *Biophys. J.*, **92**, 1732–1741.

47 Ryu, J. and Park, C.B. (2008) Solid-phase growth of nanostructures from amorphous peptide thin film: effect of

water activity and temperature. *Chem. Mater.*, **20**, 4284–4290.
48 Ryu, J. and Park, C.B. (2008) High-temperature self-assembly of peptides into vertically well-aligned nanowires by aniline vapor. *Adv. Mater.*, **20**, 3754–3758.
49 Han, T.H., Kim, J., Park, J.S., Park, C.B., Ihee, H., and Kim, S.O. (2007) Liquid crystalline peptide nanowires. *Adv. Mater.*, **19**, 3924–3927.
50 Lingenfelder, M., Tomba, G., Costantini, G., Ciacchi, L.C., De Vita, A., and Kern, K. (2007) Tracking the chiral recognition of adsorbed dipeptides at the single-molecule level. *Angew. Chem. Int. Ed.*, **46**, 4492–4495.
51 Wang, Y., Lingenfelder, M., Classen, T., Costantini, G., and Kern, K. (2007) Ordering of dipeptide chains on Cu surfaces through 2D cocrystallization. *J. Am. Chem. Soc.*, **129**, 15742–15743.
52 Kunitake, T. (1992) Synthetic bilayer membranes: molecular design, self-organization, and application. *Angew. Chem. Int. Ed.*, **31**, 709–726.
53 Ariga, K. and Kunitake, T. (1998) Molecular recognition at air–water and related interfaces: complementary hydrogen bonding and multisite interaction. *Acc. Chem. Res.*, **31**, 371–378.
54 Löwik, D.W.P.M. and van Hest, J.C. (2004) Peptide based amphiphiles. *Chem. Soc. Rev.*, **33**, 234–245.
55 Klok, H.A. (2002) Protein-inspired materials: synthetic concepts and potential applications. *Angew. Chem. Int. Ed.*, **41**, 1509–1513.
56 Zompra, A.A., Galanis, A.S., Werbitzky, O., and Albericio, F. (2009) Manufacturing peptides as active pharmaceutical ingredients. *Future Med. Chem.*, **1**, 361–377.
57 Ringsdorf, H., Schlarb, B., and Venzmer, J. (1988) Molecular architecture and function of polymeric oriented systems: models for the study of organization, surface recognition, and dynamics of biomembranes. *Angew. Chem. Int. Ed. Engl.*, **27**, 113–118.
58 Wang, K.W., Yan, X.H., Cui, Y., He, Q., and Li, J.B. (2007) Synthesis and *in vitro* behavior of multivalent cationic lipopeptide for DNA delivery and release in HeLa cells. *Bioconj. Chem.*, **18**, 1735–1738.
59 Samanta, S., Sistla, R., and Chaudhuri, A. (2010) The use of RGDGWK-lipopeptide to selectively deliver genes to mouse tumor vasculature and its complexation with p53 to inhibit tumor growth. *Biomaterials*, **31**, 1787–1797.
60 Stroumpoulis, D., Zhang, H., Rubalcava, L., Gliem, J., and Tirrell, M. (2007) Cell adhesion and growth to peptide-patterned supported lipid membranes. *Langmuir*, **23**, 3849–3856.
61 Kellam, B., De Bank, P.A., and Shakesheff, K.M. (2003) Chemical modification of mammalian cell surfaces. *Chem. Soc. Rev.*, **32**, 327–337.
62 Robson Marsden, H., Elbers, N.A., Bomans, P.H.H., Sommerdijk, N.A.J.M., and Kros, A. (2009) A reduced SNARE model for membrane fusion. *Angew. Chem. Int. Ed.*, **48**, 2330–2333.
63 Weber, T., Zemelman, B.V., McNew, J.A., Westermann, B., Gmachl, M., Parlati, F., and Rothman, J.E. (1998) SNAREpins: minimal machinery for membrane fusion. *Cell*, **92**, 759–772.
64 Cavalli, S. and Kros, A. (2008) Scope and applications of amphiphilic alkyl- and lipopeptides. *Adv. Mater.*, **20**, 627–631.
65 Cavalli, S., Albericio, F., and Kros, A. (2010) Amphiphilic peptides and their cross-disciplinary role as building blocks for nanoscience. *Chem. Soc. Rev.*, **39**, 241–263.
66 Deming, T.J. (2007) Synthetic polypeptides for biomedical applications. *Prog. Polym. Sci.*, **32**, 858–875.
67 Kricheldorf, H.R. (1987) α-*Amino Acid-N-Carboxyanhydrides and Related Materials*, Springer, New York.
68 Deming, T.J. (1999) Cobalt and iron initiators for the controlled polymerization of alpha-amino acid-*N*-carboxyanhydrides. *Macromolecules*, **32**, 4500–4502.
69 Deming, T.J. (2000) Living polymerization of α-amino acid-*N*-carboxyanhydrides. *J. Polym. Sci. Polym. Chem. Ed.*, **38**, 3011–3018.

70 Pratten, M.K., Lloyd, B., Hörpel, G., and Ringsdorf, H. (1985) Micelle forming block copolymers: pinocytosis by macrophages and interaction with model membranes. *Makromol. Chem.*, **186**, 725–733.

71 Lee, K.Y. and Mooney, D.J. (2001) Hydrogels for tissue engineering. *Chem. Rev.*, **101**, 1869–1879.

72 Peppas, N.A., Huang, Y., Torres-Lugo, M., Ward, J.H., and Zhang, J. (2000) Physicochemical, foundations and structural design of hydrogels in medicine and biology. *Annu. Rev. Biomed. Eng.*, **2**, 9–29.

73 Deming, T.J. (2005) Polypeptide hydrogels via a unique assembly mechanism. *Soft Matter*, **1**, 28–35.

74 Bellomo, E., Wyrsta, M.D., Pakstis, L., Pochan, D.J., and Deming, T.J. (2004) Stimuli responsive polypeptide vesicles via conformation specific assembly. *Nat. Mater.*, **3**, 244–248.

75 Haynie, D.T., Zhang, L., Rudra, J.S., Zhao, W., Zhong, Y., and Palath, N. (2005) Polypeptide multilayer films. *Biomacromolecules*, **6**, 2895–2913.

76 Boulmedais, F., Frisch, B., Etienne, O., Lavalle, P., Picart, C., Ogier, J., Voegel, J.C., Schaaf, P., and Egles, C. (2004) Polyelectrolyte multilayer films with pegylated polypeptides as a new type of anti-microbial protection for biomaterials. *Biomaterials*, **25**, 2003–2011.

77 Hübsch, E., Fleith, G., Fatisson, J., Labbe', P., Voegel, J.C., Schaaf, P., and Ball, V. (2005) Multivalent ion/polyelectrolyte exchange processes in exponentially growing multilayers. *Langmuir*, **21**, 3664–3669.

78 Picart, C., Mutterer, J., Richert, L., Luo, Y., Prestwich, G.D., Schaaf, P., Voegel, J.C., and Lavalle, P. (2002) Molecular basis for the explanation of the exponential growth of polyelectrolyte multilayers. *Proc. Natl Acad. Sci. USA*, **99**, 12531–12535.

79 Klitzing, R.V. and Möhwald, H. (1995) Proton concentration profile in ultrathin polyelectrolyte films. *Langmuir*, **11**, 3554–3559.

80 Yoo, D., Shiratori, S.S., and Rubner, M.F. (1998) Controlling bilayer composition and surface wettability of sequentially adsorbed multilayers of weak polyelectrolytes. *Macromolecules*, **31**, 4309–4318.

81 Rmaile, H.H. and Schlenoff, J.B. (2002) "Internal pK_a's" in polyelectrolyte multilayers: coupling protons and salt. *Langmuir*, **18**, 8263–8265.

82 Richert, L., Arntz, Y., Schaaf, P., Voegel, J.C., and Picart, C. (2004) pH dependent growth of poly(L-lysine)/poly(L-glutamic) acid multilayer films and their cell adhesion properties. *Surf. Sci.*, **570**, 13–29.

83 Zhi, Z.L. and Haynie, D.T. (2004) Direct evidence of controlled structure reorganization in a nanoorganized polypeptide multilayer thin film. *Macromolecules*, **37**, 8668–8675.

84 Zhang, L., Li, B., Zhi, Z.L., and Haynie, D.T. (2005) Perturbation of nanoscale structure of polypeptide multilayer thin films. *Langmuir*, **21**, 5439–5445.

85 Picart, C., Lavalle, P., Hubert, P., Cuisinier, F.J.G., Decher, G., Schaaf, P., and Voegel, J.C. (2001) Buildup mechanism for poly(L-lysine)/hyaluronic acid films onto a solid surface. *Langmuir*, **17**, 7414–7424.

86 Lavalle, P., Gergely, C., Cuisinier, F.J.G., Decher, G., Schaaf, P., Voegel, J.C., and Picart, C. (2002) Comparison of the structure of polyelectrolyte multilayer films exhibiting a linear and an exponential growth regime: an *in situ* atomic force microscopy study. *Macromolecules*, **35**, 4458–4465.

87 Gergely, C., Bahi, S., Szalontai, B., Flores, H., Schaaf, P., Voegel, J.C., and Cuisinier, F.J.G. (2004) Human serum albumin self-assembly on weak polyelectrolyte multilayer films structurally modified by pH changes. *Langmuir*, **20**, 5575–5582.

88 Li, B.Y., Haynie, D.T., Palath, N., and Janisch, D. (2005) Nanoscale biomimetics: fabrication and optimization of stability of peptide-based thin films. *J. Nanotechnol. Nanosci.*, **15**, 2042–2049.

89 Jessel, N.B., Atalar, F., Lavalle, P., Mutterer, J., Decher, G., Schaaf, P., Voegel, J.C., and Ogier, J. (2003) Bioactive coatings based on a

polyelectrolyte multilayer architecture functionalized by embedded proteins. *Adv. Mater.*, **15**, 692–695.

90 Haynie, D.T., Palath, N., Liu, Y., Li, B., and Pargaonkar, N. (2005) Biomimetic nanostructured materials: inherent reversible stabilization of polypeptide microcapsules. *Langmuir*, **21**, 1136–1138.

91 Boulmedais, F., Schwinté, P., Gergely, C., Voegel, J.C., and Schaaf, P. (2002) Secondary structure of polypeptide multilayer films: an example of locally ordered polyelectrolyte multilayers. *Langmuir*, **18**, 4523–4525.

93 Boulmedais, F., Bozonnet, M., Schwinté, P., Voegel, J.C., and Schaaf, P. (2003) Multilayered polypeptide films: secondary structures and effect of various stresses. *Langmuir*, **19**, 9873–9882.

92 Haynie, D.T., Balkundi, S., Palath, N., Chakravarthula, K., and Dave, K. (2004) Polypeptide multilayer films: role of molecular structure and charge. *Langmuir*, **20**, 4540–4547.

94 Debreczeny, M., Ball, V., Boulmedais, F., Szalontai, B., Voegel, J.C., and Schaaf, P. (2003) Multilayers built from two component polyanions and single component polycation solutions: a way to engineer films with desired secondary structure. *J. Phys. Chem. B.*, **107**, 12734–12739.

95 Boulmedais, F., Ball, V., Schwinte, P., Frisch, B., Schaaf, P., and Voegel, J.C. (2003) Buildup of exponentially growing multilayer polypeptide films with internal secondary structure. *Langmuir*, **19**, 440–445.

96 Chluba, J., Voegel, J.C., Decher, G., Erbacher, P., Schaaf, P., and Ogier, J. (2001) Peptide hormone covalently bound to polyelectrolytes and embedded into multilayer architectures conserving full biological activity. *Biomacromolecules*, **2**, 800–805.

97 Halthur, T.J., Claesson, P.M., and Elofsson, U.M. (2004) Stability of polypeptide multilayers as studied by *in situ* ellipsometry: effects of drying and post-buildup changes in temperature and pH. *J. Am. Chem. Soc.*, **126**, 17009–17015.

98 Halthur, T.J. and Elofsson, U.M. (2004) Multilayers of charged polypeptides as studied by *in situ* ellipsometry and quartz crystal microbalance with dissipation. *Langmuir*, **20**, 1739–1745.

99 Li, B. and Haynie, D.T. (2004) Multilayer biomimetics: reversible covalent stabilization of a nanostructured biofilm. *Biomacromolecules*, **5**, 1667–1670.

100 Aoki, T., Tomizawa, S., and Oikawa, E. (1995) Enantioselective permeation through poly(gamma-[3-(pentamethyldisiloxanyl)propyl]-L-glutamate) membranes. *J. Membr. Sci.*, **99**, 117–125.

101 Rmaile, H.H. and Schlenoff, J.B. (2003) Optically active polyelectrolyte multilayers as membranes for chiral separations. *J. Am. Chem. Soc.*, **125**, 6602–6603.

102 Etienne, O., Picart, C., Taddei, C., Haikel, Y., Dimarcq, J.L., Schaaf, P., Voegel, J.C., Ogier, J.A., and Engles, C. (2004) Multilayer polyelectrolyte films functionalized by insertion of defensin: a new approach to protection of implants from bacterial colonization. *Antimicrob. Agents Chemother.*, **48**, 3662–3669.

103 Balasubramani, M., Ravikumar, T., and Babu, M. (2001) Skin substitutes: a review. *Burns*, **27**, 534–544.

104 Picart, C., Elkaim, R., Richert, L., Audoin, F., Arntz, Y., Cardoso, M.D.S., Schaaf, P., Voegel, J.C., and Frisch, B. (2005) Primary cell adhesion on RGD-functionalized and covalently crosslinked thin polyelectrolyte multilayer films. *Adv. Funct. Mater.*, **15**, 83–94.

105 Chang, T.M.S. (2005) Therapeutic applications of polymeric artificial cells. *Nat. Rev. Drug Discov.*, **4**, 221–235.

106 Jessel, N.B., Schwinté, P., Donahue, R., Lavalle, P., Boulmedais, F., Szalontai, B., Voegel, J.C., and Ogier, J. (2004) Pyridylamino-beta-cyclodextrin as a molecular chaperone for lipopolysaccharide embedded in a multilayered polyelectrolyte architecture. *Adv. Funct. Mater.*, **14**, 963–969.

107 He, Q., Tian, Y., Cui, Y., Möhwald, H., and Li, J.B. (2008) Layer-by-layer

assembly of magnetic polypeptide nanotubes as a DNA carrier. *J. Mater. Chem.*, **18**, 748–754.

108 Yu, A.M., Gentle, I., Lu, G.Q., and Caruso, F. (2006) Nanoassembly of biocompatible microcapsules for urease encapsulation and their use as biomimetic reactors. *Chem. Commun.*, 2150–2152.

110 Zhao, X.J. and Zhang, S.G. (2006) Molecular designer self-assembling peptides. *Chem. Soc. Rev.*, **35**, 1105–1110.

109 Zhao, X.J. (2009) Design of self-assembling surfactant-like peptides and their applications. *Curr. Opin. Colloid Interface Sci.*, **14**, 340–348.

111 Ghadiri, M.R., Granja, J.R., Milligan, R.A., McRee, D.E., and Khazanovich, N. (1993) Self-assembling organic nanotubes based on a cyclic peptide architecture. *Nature*, **366**, 324–327.

112 Hartgerink, J.D., Granja, J.R., Milligan, R.A., and Ghadiri, M.R. (1996) Self-assembling peptide nanotubes. *J. Am. Chem. Soc.*, **118**, 43–50.

113 Seebach, D., Matthews, J.L., Meden, A., Wessels, T., Naerlocher, C., and Mccusker, L.B. (1997) Cyclo-β-peptides: structure and tubular stacking of cyclic tetramers of 3-aminobutanoic acid as determined from powder diffraction data. *Helv. Chim. Acta*, **80**, 173–182.

114 Clark, T.D., Buriak, J.M., Kobayashi, K., Isler, M.P., McRee, D.E., and Ghadiri, M.R. (1998) Cylindrical β-sheet peptide assemblies. *J. Am. Chem. Soc.*, **120**, 8949–8962.

115 Amorin, M., Castedo, L., and Granja, J.R. (2003) New cyclic peptide assemblies with hydrophobic cavities: the structural and thermodynamic basis of a new class of peptide nanotubes. *J. Am. Chem. Soc.*, **125**, 2844–2845.

116 Semetey, V., Didierjean, C., Briand, J.P., Aubry, A., and Guichard, G. (2002) Self-assembling organic nanotubes from enantiopure cyclo-*N,N'*-linked oligoureas: design, synthesis, and crystal structure. *Angew. Chem. Int. Ed.*, **41**, 1895–1898.

117 Ranganathan, D., Lakshmi, C., and Karle, I.L. (1999) Hydrogen-bonded self-assembled peptide nanotubes from cystine-based macrocyclic bisureas. *J. Am. Chem. Soc.*, **121**, 6103–6107.

118 Horne, W.S., Stout, C.D., and Ghadiri, M.R. (2003) A heterocyclic peptide nanotube. *J. Am. Chem. Soc.*, **125**, 9372–9376.

119 Woolfson, D.N. and Ryadnov, M.G. (2006) Peptide-based fibrous biomaterials: some things old, new and borrowed. *Curr. Opin. Chem. Biol.*, **10**, 559–567.

120 Ryadnov, M.G. and Woolfson, D.N. (2007) Self-assembled templates for polypeptide synthesis. *J. Am. Chem. Soc.*, **129**, 14074–14081.

121 Pandya, M.J., Spooner, G.M., Sunde, M., Thorpe, J.R., Rodger, A., and Woolfson, D.N. (2000) Sticky-end assembly of a designed peptide fiber provides insight into protein fibrillogenesis. *Biochemistry*, **39**, 8728–8734.

122 Papapostolou, D., Bromley, E.H.C., Bano, C., and Woolfson, D.N. (2008) Electrostatic control of thickness and stiffness in a designed protein fiber. *J. Am. Chem. Soc.*, **130**, 5124–5130.

123 Mason, J.M. and Arndt, K.M. (2004) Coiled coil domains: stability, specificity, and biological implications. *ChemBioChem*, **5**, 170–176.

124 Hwang, J.J., Iyer, S.N., Li, L.S., Claussen, R., Harrington, D.A., and Stupp, S.I. (2002) Self-assembling biomaterials: liquid crystal phases of cholesteryl oligo(L-lactic acid) and their interactions with cells. *Proc. Natl. Acad. Sci. USA*, **99**, 9662–9667.

125 Klok, H.A., Hwang, J.J., Hartgerink, J.D., and Stupp, S.I. (2002) Self-assembling biomaterials: L-lysine-dendron-substituted cholesteryl-(L-lactic acid)$_n$. *Macromolecules*, **35**, 6101–6111.

126 Klok, H.A., Hwang, J.J., Iyer, S.N., and Stupp, S.I. (2002) Cholesteryl-(L-lactic acid)$_n$ building blocks for self-assembling biomaterials. *Macromolecules*, **35**, 746–759.

127 Hartgerink, J.D., Beniash, E., and Stupp, S.I. (2002) Peptide-amphiphile nanofibers: a versatile scaffold for the preparation of self-assembling materials. *Proc. Natl. Acad. Sci. USA*, **99**, 5133–5138.

128 Cui, H.G., Webber, M.J., and Stupp, S.I. (2010) Self-assembly of peptide amphiphiles: from molecules to nanostructures to biomaterials. *Biopolymers*, **94**, 1–18.

129 Paramonov, S.E., Jun, H.W., and Hartgerink, J.D. (2006) Self-assembly of peptide-amphiphile nanofibers: the roles of hydrogen bonding and amphiphilic packing. *J. Am. Chem. Soc.*, **128**, 7291–7298.

130 Behanna, H.A., Donners, J., Gordon, A.C., and Stupp, S.I. (2005) Coassembly of amphiphiles with opposite peptide polarities into nanofibers. *J. Am. Chem. Soc.*, **127**, 1193–1200.

131 Niece, K.L., Hartgerink, J.D., Donners, J.J.J.M., and Stupp, S.I. (2003) Self-assembly combining two bioactive peptide–amphiphile molecules into nanofibers by electrostatic attraction. *J. Am. Chem. Soc.*, **125**, 7146–7147.

132 Hsu, L., Cvetanovich, G.L., and Stupp, S.I. (2008) Peptide amphiphile nanofibers with conjugated polydiacetylene backbones in their core. *J. Am. Chem. Soc.*, **130**, 3892–3899.

133 Guler, M.O., Hsu, L., Soukasene, S., Harrington, D.A., Hulvat, J.F., and Stupp, S.I. (2006) Presentation of RGDS epitopes on self-assembled nanofibers of branched peptide amphiphiles. *Biomacromolecules*, **7**, 1855–1863.

134 Tovar, J.D., Claussen, R.C., and Stupp, S.I. (2005) Probing the interior of peptide amphiphile supramolecular aggregates. *J. Am. Chem. Soc.*, **127**, 7337–7345.

135 Guler, M.O., Claussen, R.C., and Stupp, S.I. (2005) Encapsulation of pyrene within self-assembled peptide amphiphile nanofibers. *J. Mater. Chem.*, **15**, 4507–4512.

136 Pashuck, E.T., Cui, H.G., and Stupp, S.I. (2010) Tuning supramolecular rigidity of peptide fibers through molecular structure. *J. Am. Chem. Soc.*, **132**, 6041–6046.

137 Chen, C.L. and Rosi, N.L. (2010) Peptide-based methods for the preparation of nanostructured inorganic materials. *Angew. Chem. Int. Ed.*, **49**, 1924–1942.

138 Fu, X.Y., Wang, Y., Huang, L.X., Sha, Y.L., Gui, L.L., Lai, L.H., and Tang, Y.Q. (2003) Assemblies of metal nanoparticles and self-assembled peptide fibrils – formation of double helical and single-chain arrays of metal nanoparticles. *Adv. Mater.*, **15**, 902–906.

139 Sharma, N., Top, A., Kiick, K.L., and Pochan, D.J. (2009) One-dimensional gold nanoparticle arrays by electrostatically directed organization using polypeptide self-assembly. *Angew. Chem. Int. Ed.*, **48**, 7078–7082.

140 Li, L.S. and Stupp, S.I. (2005) One-dimensional assembly of lipophilic inorganic nanoparticles templated by peptide-based nanofibers with binding functionalities. *Angew. Chem. Int. Ed.*, **44**, 1833–1836.

141 Lamm, M.S., Sharma, N., Rajagopal, K., Beyer, F.L., Schneider, J.P., and Pochan, D.J. (2008) Laterally spaced linear nanoparticle arrays templated by laminated β-sheet fibrils. *Adv. Mater.*, **20**, 447–551.

142 Euliss, L.E., Grancharov, S.G., Brien, S.O., Deming T.J., Stucky, G.D., Murray, C.B., and Held, G.A. (2003) Cooperative assembly of magnetic nanoparticles and block copolypeptides in aqueous media. *Nano Lett.*, **3**, 1489–1493.

143 Wong, M.S., Cha, J.N., Choi, K.S., Deming, T.J., and Stucky, G.D. (2002) Assembly of nanoparticles into hollow spheres using block copolypeptides. *Nano Lett.*, **2**, 583–587.

144 Yan, X.H., Cui, Y., Qi, W., Su, Y., Yang, Y., He, Q., and Li, J.B. (2008) Self-assembly of peptide-based colloids containing lipophilic nanocrystals. *Small*, **4**, 1687–1693.

145 Chen, C.L., Zhang, P.J., and Rosi, N.L. (2008) A new peptide-based method for the design and synthesis of nanoparticle superstructures: construction of highly ordered gold nanoparticle double helices. *J. Am. Chem. Soc.*, **130**, 13555–13557.

146 Behrens, S.S. (2008) Synthesis of inorganic nanomaterials mediated by protein assemblies. *J. Mater. Chem.*, **18**, 3788–3798.

147 Dickerson, M.B., Sandhage, K.H., and Naik, R.R. (2008) Protein- and peptide-directed syntheses of inorganic materials. *Chem. Rev.*, **108**, 4935–4978.

148 Dickerson, M.B., Naik, R.R., Stone, M.O., Cai, Y., and Sandhage, K.H. (2004) Identification of peptides that promote the rapid precipitation of germania nanoparticle networks via use of a peptide display library. *Chem. Commun.*, 1776–1777.

149 Naik, R.R., Jones, S.E., Murray, C.J., McAuliffe, J.C., Vaia, R.A., and Stone, M.O. (2004) Peptide templates for nanoparticle synthesis derived from polymerase chain reaction-driven phage display. *Adv. Funct. Mater.*, **14**, 25–30.

150 Volkmer, D., Fricke, M., Huber, T., and Sewald, N. (2004) Acidic peptides acting as growth modifiers of calcite crystals. *Chem. Commun.*, 1872–1873.

151 Djalali, R., Chen, Y.F., and Matsui, H. (2003) Nanocrystal growth on nanotube controlled by conformations and charges of sequenced peptide templates. *J. Am. Chem. Soc.*, **125**, 5873–5879.

152 Banerjee, I.A., Yu, L.T., and Matsui, H. (2003) Cu nanocrystal growth on peptide nanotubes by biomineralization: size control of Cu nanocrystals by tuning peptide conformation. *Proc. Natl. Acad. Sci. USA*, **100**, 14678–14682.

153 Yu, L.T., Banerjee, I.A., and Matsui, H. (2003) Direct growth of shape-controlled nanocrystals on nanotubes via biological recognition. *J. Am. Chem. Soc.*, **125**, 14837–14840.

154 Tas, A.C. (2000) Synthesis of biomimetic Ca-hydroxyapatite powders at 37 degrees C in synthetic body fluids. *Biomaterials*, **21**, 1429–1438.

155 Liu, D.M., Troczynski, T., and Tseng, W.J. (2001) Water-based sol-gel synthesis of hydroxyapatite: process development. *Biomaterials*, **22**, 1721–1730.

156 Palmer, L.C., Newcomb, C.J., Kaltz, S.R., Spoerke, E.D., and Stupp, S.I. (2008) Biomimetic systems for hydroxyapatite mineralization inspired by bone and enamel. *Chem. Rev.*, **108**, 4754–4783.

157 Spoerke, E.D., Anthony, S.G., and Stupp, S.I. (2009) Enzyme directed templating of artificial bone mineral. *Adv. Mater.*, **21**, 425–430.

158 Long, D., Burkholder, E., and Cronin, L. (2007) Polyoxometalate clusters, nanostructures and materials – from self assembly to designer materials and devices. *Chem. Soc. Rev.*, **36**, 105–121.

159 Yan, X.H., Zhu, P.L., Fei, J.B., Möhwald, H., and Li, J.B. (2010) Self-assembly of peptide–inorganic hybrid spheres for adaptive encapsulation of guests. *Adv. Mater.*, **22**, 1283–1287.

160 Abbott, A. (2003) Cell culture: biology's new dimension. *Nature*, **424**, 870–872.

161 Zhang, S., Holmes, T., Lockshin, C., and Rich, A. (1993) Spontaneous assembly of a self-complementary oligopeptide to form a stable macroscopic membrane. *Proc. Natl. Acad. Sci. USA*, **90**, 3334–3338.

162 Zhang, S., Holmes, T., DiPersio, M., Hynes, R.O., Su, X., and Rich, A. (1995) Self-complementary oligopeptide matrices support mammalian cell attachment. *Biomaterials*, **1**, 1385–1390.

163 Holmes, T., Delacalle, S., Su, X., Rich, A., and Zhang, S. (2000) Extensive neurite outgrowth and active neuronal synapses on peptides scaffolds. *Proc. Natl. Acad. Sci. USA*, **97**, 6728–6733.

164 Kisiday, J., Jin, M., Kurz, B., Hung, H., Semino, C., Zhang, S., and Grodzinsky, A.J. (2002) Self-assembling peptide hydrogel fosters chondrocyte extracellular matrix production and cell division: implications for cartilage tissue repair. *Proc. Natl. Acad. Sci. USA*, **99**, 9996–10001.

165 Silva, G.A., Czeisler, C., Niece, K.L., Beniash, E., Harrington, D.A., Kessler, J.A., and Stupp, S.I. (2004) Selective differentiation of neural progenitor cells by high-epitope density nanofibers. *Science*, **303**, 1352–1355.

166 Mata, A., Hsu, L., Capito, R., Aparicio, C., Henrikson, K., and Stupp, S.I. (2009) Micropatterning of bioactive self-assembling gels. *Soft Matter*, **5**, 1228–1236.

167 Hung, A.M. and Stupp, S.I. (2007) Simultaneous self-assembly, orientation, and patterning of peptide–

amphiphile nanofibers by soft lithography. *Nano Lett.*, **7**, 1165–1171.
168 Hung, A.M. and Stupp, S.I. (2009) Understanding factors affecting alignment of self-assembling nanofibers patterned by sonication-assisted solution embossing. *Langmuir*, **25**, 7084–7089.
169 Beniash, E., Hartgerink, J.D., Storrie, H., Stendahl, J.C., and Stupp, S.I. (2005) Self-assembling peptide amphiphile nanofiber matrices for cell entrapment. *Acta Biomater.*, **1**, 387–397.
170 Zhou, M., Smith, A.M., Das, A.K., Hodson, N.W., Collins, R.F., Ulijn, R.V., and Gough, J.E. (2009) Self-assembled peptide-based hydrogels as scaffolds for anchorage dependent cells. *Biomaterials*, **30**, 2523–2530.
171 Xu, Y.H. and Szoka, F.C. (1996) Mechanism of DNA release from cationic liposome/DNA complexes used in cell transfection. *Biochemistry*, **35**, 5616–5623.
172 Prata, C.A.H., Zhao, Y.X., Barthelemy, P., Li, Y., Luo, D., McIntosh, T.J., Lee, S.J., and Grinstaff, M.W. (2004) Charge reversal amphiphiles for gene delivery. *J. Am. Chem. Soc.*, **126**, 12196–12197.
173 Holowka, E.P., Sun, V.Z., Kamei, D.T., and Deming, T.J. (2007) Polyarginine segments in block copolypeptides drive both vesicular assembly and intracellular delivery. *Nat. Mater.*, **6**, 52–57.
174 Calnan, B.J., Tidor, B., Biancalana, S., Hudson, D., and Frankel, A.D. (1991) Arginine-mediated RNA recognition: the arginine fork. *Science*, **252**, 1167–1171.
175 Mitchell, D.J., Kim, D.T., Steinman, L., Fathman, C.G., and Rothbard, J.B. (2000) Polyarginine enters cells more efficiently than other polycationic homopolymers. *J. Peptide Res.*, **56**, 318–325.
176 Rothbard, J.B., Garlington, S., Lin, Q., Kirschberg, T., Kreider, E., McGrane, P.L., Wender, P.A., and Khavari, P.A. (2000) Conjugation of arginine oligomers to cyclosporin A facilitates topical delivery and inhibition of inflammation. *Nat. Med.*, **6**, 1253–1257.
177 Missirlis, D., Khant, H., and Tirrell, M. (2009) Mechanisms of peptide amphiphile internalization by SJSA-1 cells in vitro. *Biochemistry*, **48**, 3304–3314.
178 Yemini, M., Reches, M., Rishpon, J., and Gazit, E. (2005) Novel electrochemical biosensing platform using self-assembled peptide nanotubes. *Nano Lett.*, **5**, 183–186.
179 Yemini, M., Reches, M., Gazit, E., and Rishpon, J. (2005) Peptide nanotube-modified electrodes for enzyme biosensor. *Anal. Chem.*, **77**, 5155–5159.
180 Carny, O., Shalev, D.E., and Gazit, E. (2006) Fabrication of coaxial metal nanocables using a self-assembled peptide nanotube scaffold. *Nano Lett.*, **6**, 1594–1597.
181 Pouget, E., Dujardin, E., Cavalier, A., Moreac, A., Valéry, C., Marchi-Artzner, V., Weiss, T., Renault, A., Paternostre, M., and Artzner, F. (2007) Hierarchical architectures by synergy between dynamical template self-assembly and biomineralization. *Nat. Mater.*, **6**, 434–439.
182 Han, T.H., Oh, J.K., Park, J.S., Kwon, S.H., Kim, S.W., and Kim, S.O. (2009) Highly entangled hollow TiO_2 nanoribbons templating diphenylalanine assembly. *J. Mater. Chem.*, **19**, 3512–3516.
183 Ryu, J. and Park, C.B. (2009) Synthesis of diphenylalanine–polyaniline core–shell conducting nanowires by peptide self-assembly. *Angew. Chem. Int. Ed.*, **48**, 4820–4823.
184 Lee, J.S., Ryu, J., and Park, C.B. (2009) Bio-inspired fabrication of superhydrophobic surfaces through peptide self-assembly. *Soft Matter*, **5**, 2717–2720.

Glossary

6-CF	6-carboxyfluorescein
AD	neurodegenerative, dementia-inducing disorder characterized mainly by amyloid deposits surrounding dying neurons, neurofibrillar tangles, and cerebrovascular angiopathies
ADH	ethanol dehydrogenase
ADP	adenosine diphosphate
AFM	atomic force microscopy
ALD	atomic layer deposition
A-phase	aggregate phase
ATP	adenosine triphosphate
ATRP	atom transfer radical polymerization
AUT	11-aminoundecanethiol
Aβ	amyloid β
BAM	Brewster angle microscopy
BIBB	bromoisobutyryl bromide – a surface initiator for polymerization
CAC	critical aggregation concentration
CD	circular dichroism
CDPNT	cationic dipeptide nanotube
CFF	cysteine–diphenylalanine tripeptide
CL	chemical lithography
CLSM	confocal laser scanning microscopy
C–P–Q	carotenoid–porphyrin–quinone
CR	Congo Red
CS$_2$	carbon disulfide
CTVC	critical tube vesicle concentration
DDAB	dimethyldioctadecylammonium bromide
DDT	1-dodecanethiol
DHP	dihexadecylphosphate, sodium salt
DHPE	dihexadecanoylphosphatidylethanolamine
DLPE	dilauroylphosphatidylethanolamine
DLS	dynamic light scattering
DMPA	L-dimyristoyl-*sn*-glycero-3-phosphatidic acid

Molecular Assembly of Biomimetic Systems. Junbai Li, Qiang He, and Xuehai Yan
© 2011 WILEY-VCH Verlag GmbH & Co. KGaA, Weinheim
ISBN: 978-3-527-32542-9

DMPC	dimyristoylphosphatidylcholine
DMPE	dimyristoylphosphatidylethanolamine
DODAB	didodecyldimethylammonium bromide
DOPA	dioleoylphosphatidic acid
DOPE	1,2-dioleoyl-*sn*-glycero-3-phosphatidylethanolamine
DPG	dipalmitoylglycerol
DPN	dip-pen nanolithography
DPPA	dipalmitoylphosphatidic acid
DSC	differential scanning calorimetry
ECM	extracellular matrix
EDTA	ethylenediamine tetraacetate
EDX	energy-dispersive X-ray spectrum
ESEM	environmental scanning electron microscopy
EtBr	ethidium bromide
FF	diphenylalanine
FITC	fluorescein isothiocyanate
FNT	diphenylalanine nanotube
F_oF_1-ATPase	class of integral membrane proteins; this enzyme is comprised of two separate domains: F_o, the hydrophobic membrane-bound portion, which is responsible for proton translocation, and the hydrophilic membrane-bound portion F_1, which is responsible for ATP hydrolysis or synthesis.
FRAP	fluorescence recovery after photobleaching
FTIR	Fourier transform infrared spectroscopy
GA	glutaraldehyde
GFP	Green Fluorescent Protein
GIXRD	grazing incidence X-ray diffraction
GOD	glucose oxidase
GPI	glycosylphophatidylinositol
HA	hydroxyapatite
Hb	hemoglobin
HFIP	1,1,1,3,3-hexafluoro-2-propanol
HRTEM	high-resolution transmission electron microscopy
HSA	human serum albumin
IgG	immunoglobulin G
I-phase	isotropic phase
IS	impedance spectroscopy
Kinesin	motor protein capable of moving along microtubule cables powered by the hydrolysis of ATP; many cellular functions including mitosis, meiosis, and cargo transport are concerned with the active movement of kinesins
LB	Langmuir–Blodgett
LS	Langmuir–Schäfer
LbL assembly	layer-by-layer assembly – a technique where multilayer structures are formed in a layer-by-layer stacking fashion,

	notably by the adsorption of alternatively polyions on arbitrarily shaped objects
LC phase	liquid-condensed phase
LCST	lower critical solution temperature
L-DPPC	L-α-dipalmitoylphosphatidylcholine
LE phase	liquid-expanded phase
Liposome	artificial vesicle composed of lipid bilayers
LMOG	low-molecular-mass organogelator
LZ	leucine zipper
MCL	multivalent cationic lipopeptide
MF	melamine formaldehyde
MS	mesoporous silica
MSC	mesenchymal stem cell
MTT	3-(4,5-dimethylthiazolyl-2)-2,5-diphenyltetrazolium bromide
NBD	7-nitrobenz-2-oxa-1,3-diazol-4-yl
NBD-DPPE	NBD-labeled 1,2-dipalmitoyl-sn-glycero-3-phosphoethanolamine
NBD-PC	NBD-labeled phosphatidylcholine
NBT	4′-nitro-1,1′-biphenyl-4-thiol
NCA	α-amino acid-N-carboxyanhydride
Ni-NTA	Ni^{2+}-nitrilotriacetic acid
NIPAM	N-isopropylacrylamide
NMP	N-methyl-2-pyrrolidone
NR	neutron reflectometry
NSF	N-ethylmaleimide-sensitive factor
PA	phosphatidic acid
PAA	poly(acrylic acid) (sodium salt)
PAH	poly(allylamine hydrochloride)
PANI	polyaniline
PAT	profile analysis tensiometry technique
PB1124	10-tetradecyloxymethy-3,6,9,12-tetraoxahexacosyl 2-acetamido-2-deoxy-β-D-glucopyranoside
PC	phosphatidylcholine
PCR	polymerase chain reaction
PDDA	poly(diallyldimethylammonium chloride)
PDMS	polydimethylsiloxane
PEC	peptide encapsulated cluster
PEG	polyethylene glycol
PEI	polyethyleneimine
PGA	poly-L-glutamic acid
P_i	phosphate ion
PL	photoluminescent
PLA_1	phospholipase A_1
PLA_2	phospholipase A_2
PLC	phospholipase C
PLD	phospholipase D

PLL	poly-L-lysine hydrochloride
PM-IRRAS	polarization-modulated external infrared reflection absorption
PNIPAM	poly(N-isopropylacrylamide)
POM	polyoxometalate
proteoliposome	liposome incorporating purified membrane proteins
PS	phosphatidylserine
PSS	poly(styrenesulfonate) sodium salt
PTA	phosphotungstic acid
PTCDA	3,4,9,10-perylenetetracarboxylicdianhydride
PTD	protein-transduction domains
Pyranine	8-hydroxy-1,3,6-pyrenetrisulfonate
QCM	quartz crystal microbalance
QD	quantum dot
R6G	rhodamine 6G
RGD	arginine–glycine–aspartic acid – tripeptide motif capable of specifically recognizing integrins
RLP	rubella-like particle
ROMP	ring-opening metathesis polymerization
SAM	self-assembled monolayer
SEM	scanning electron microscopy
SI-ATRP	surface-initiated atom-transfer radical polymerization
SIP	surface-initiated polymerization
SNARE	soluble NSF attachment protein receptor
SPLS	single particle light scattering
SPM	scanning probe microscopy
ssDNA	single-stranded DNA
STM	scanning tunneling microscopy
TEM	transmission electron microscopy
TGA	thermogravimetric analysis
TiO_2	titanium dioxide
TPA	terephthalic acid
Triton X-100	polyethylene glycol *p*-(1,1,3,3-tetramethylbutyl)-phenyl ether, a nonionic surfactant
UV	ultraviolet-visible spectroscopy
XPS	X-ray photoelectron spectroscopy
XRD	X-ray diffraction
β-CA	β-casein
β-LG	β-lactoglobulin
μCP	microcontact printing
π–A isotherm	surface pressure (π) as a function of the molecular area (A)

Index

Π–A isotherm 9, 10

a
adaptive hybrid supramolecular networks 164, 165
affinity-capture assays 108, 109
aggregate phase (A-phase) 134
alternating site hypothesis 65
Alzheimer's Aβ polypeptide 130, 131
Alzheimer's disease 136
α-amino acid–N-carboxyanhydrides (NCAs) 146, 147
11-aminoundecanethiol (AUT) 123
amphiphilic peptides 150–156
arginine–glycine–aspartic acid (RGD) sequence 143, 144
aromatic dipeptides, nanostructure fabrication 130, 131
– nanofibrils and ribbons 136–139
– nanotubes, nanotube arrays, and vesicles 131–136
– nanowires 139, 140
– ordered molecular chains on solid surfaces 140
artificial photosynthesis 74–76
atom transfer radical polymerization (ATRP) 115
atomic force microscopy (AFM) 19
atomic layer deposition (ALD) 171

b
bacteriorhodopsin 71–74
bead geometry 93, 94
– layer-by-layer (LbL) assembly of kinesin–microtubule-driven systems 97, 98
binding charge hypothesis 65
bioimaging 169, 170
biomimetic interface 103, 104

biomineralization 160–164
biomolecule patterning 104, 105, 123, 124
– covalent immobilization of lipid monolayers 110–113
– covalent immobilization of proteins 108–110
– electrostatic immobilization of proteins 105
– – lipid-modified HSA for *E. coli* recognition 107
– – lipid-modified HSA for targeted recognition 105–107
– polymer brush patterns for biomedical applications 114
– – fabrication of complex brush gradients 118–123
– – thermosensitve patterns for cell adhesion 114–118
biosensors 170
biotin–streptavidin recognition 93
bone biomimetic mineralization 163, 164
Brewster angle microscopy (BAM) 22, 24, 25

c
carotenoid–porphyrin–quinone (C–P–Q) triad in artificial photosynthesis 74–76
β-casein (β-CA) 15, 16
– skin-like films on curved surfaces 16, 17
cell culture, three-dimensional scaffolds 165–167
cell membranes
– functions 7
chondrocyte cells 167
circular dichroism (CD) 133
compartmentalization of cells 41
confocal scanning laser microscopy (CLSM) 29, 49

Molecular Assembly of Biomimetic Systems. Junbai Li, Qiang He, and Xuehai Yan
© 2011 WILEY-VCH Verlag GmbH & Co. KGaA, Weinheim
ISBN: 978-3-527-32542-9

copper nanocrystals 161
covalent immobilization of lipid monolayers 110–113
covalent immobilization of proteins 108–110
crital aggregate concentration (CAC) 13
critical tube vesicle concentration (CVTC) 134, 135
cysteine–diphenylalanine tripeptide (CFF) 135

d

Delaunay rotationally symmetric hypersurface 135
diabetes, type II 136
differential scanning calorimetry (DSC) 49
dihexadecyl phosphate (DHP) 51, 52
dilauroylphosphatidylethanolamine (DLPE) 57
dimethyldioctadecylammonium bromide (DDAB) 51
L-α-dimyristoylphosphatidic acid (DMPA) 105, 106
– SAMs 112
dimyristoylphosphatidylethanolamine (DMPE)
– PAT analysis 10–14
dipalmitoylphosphatidic acid (DPPA) 19, 20
L-dipalmitoylphosphatidylcholine (L-DPPC) 9
– composite layers with β-LG
–– dynamic adsorption mechanism 18
– PAT analysis 10–14
dipalmitoylphosphatidylethanolamine (DPPE) 52
– composite layers with β-LG
–– dynamic adsorption mechanism 18, 19
diphenylalanine dipeptide (FF) 130, 131
– nanofibrils and ribbons 136–139
– nanotubes, nanotube arrays, and vesicles 131–136
– nanowires 139, 140
– ordered molecular chains on solid surfaces 140
drug delivery 167, 168
dynamic light scattering (DLS) 160

e

electron beam lithography (EBL) 110
electron dose 119
electrostatic immobilization of proteins 105
– lipid-modified HSA for E. coli recognition 107
– lipid-modified HAS for targeted recognition 105–107
Escherichia coli
– recognition by lipid-modified HSA 107
ethidium bromide (EtBr) 142, 143
N-ethylmaleimide-sensitive factor (NSF) 144, 145

f

F_oF_1–ATP synthase-based systems 63, 85
– ATP biosynthesis from microcapsules
–– proton gradient generation 76–78
–– proton gradients from GOD capsules 80–82
–– proton gradients from oxidative hydrolysis of glucoses 78–80
–– reassembly in polymerosomes 82–85
– direct observation 67–70
– reconstitution in cellular mimic structures 70, 71
–– bacteriorhodopsin 71–74
–– liposome incorporation 71
–– proton gradients 74–76
– rotary molecular motor 63–64
–– structure of $H^+F_oF_1$-ATPase 64–67
FF nanotubes (FNTs) 131, 132, 170
9-fluorenylmethoxycarbonyl (Fmoc)-protected diphenylalanine (Fmoc-FF) 136
fluorescence recovery after photobleaching (PRAP) 27
Frumkin isotherm 13, 14

g

gene delivery 167, 168
gliding geometry 94
– layer-by-layer (LbL) assembly of kinesin–microtubule-driven systems 98–100
glucose oxidase (GOD) 46
– proton gradients 78–80
–– capsules 80–82
gold nanoparticles 157, 161
gradient surfaces 119
grazing incidence X-ray diffraction (GIXRD) 22
Green Fluorescent Protein (GFP) 107

h

$H^+F_oF_1$–ATPase
– direct observation 67–70
– structure 64–67
haemoglobin (Hb) 53, 54
Hansen–Joos equation 13, 15
human serum albumin (HSA) 15, 16
– isoelectric point 105

- lipid-modification for *E. coli* recognition   107
- lipid-modification for targeted recognition   105–107
- skin-like films on curved surfaces   16, 17
Huntington disease   136
hydrogels   147, 155
hydroxyapatite (HA)   162–164

i

inorganic nanoparticles attached on peptide-based scaffolds   156–160
integrin-binding extracellular protein (ECM)   143, 165, 166
isotropic phase (I-phase)   134

k

kinesin-1   92
kinesin–microtubule-driven systems   91, 92, 100
- active biomimetic systems   93
-- bead geometry   93, 94
-- gliding geometry   94
-- transport direction and distance   95, 96
- active transport   92, 93
- layer-by-layer (LbL) assembly
-- fabrication in beaded geometry   97, 98
-- fabrication in gliding geometry   98–100
-- hollow microcapsules   96, 97

l

β-lactoglobulin (β-LG)   15, 16
- composite layers with DPPC
-- dynamic adsorption mechanism   18
- composite layers with DPPE
-- dynamic adsorption mechanism   18, 19
- isoelectric point   19
- skin-like films on curved surfaces   16, 17
lag–burst enzymatic cleavage   23
Langmuir adsorption isotherm   13
Langmuir method   10
Langmuir–Blodgett (LB) balance   11
Langmuir–Blodgett films   112
Langmuir–Schafer technique   27, 28
layer-by-layer (LbL) assembly   41, 42, 58
- kinesin–microtubule-driven systems
-- fabrication in beaded geometry   97, 98
-- fabrication in gliding geometry   98–100
-- hollow microcapsules   96, 97
- microcapsules   42–45
- microcapsules, biomimetic
-- permeation and mechanical properties   45–47
- polypeptide multilayer films   148, 149

lipid monolayers   30, 31
- covalent immobilization   110–113
- modeling hydrolysis *in vitro*   20, 21
-- PLA_2 phospholipase   21–23
-- PLC phospholipase   23, 24
-- PLD phospholipase   24, 25
- phospholipid monolayers
-- air/water interface   8–10
-- interfacial behavior   10–14
-- oil/water interface   10
- phospholipid/protein composite layers
-- interfacial behavior   17–20
- polyelectrolyte-supported lipid bilayers   25–27
-- multilayers on curved surfaces   28–30
-- multilayers on planar surfaces   27, 28
- protein layers
-- oil/water interface   15–17
lipopeptides   140–145
liposomes   25
- F_oF_1–ATPase incorporation   71
lotus-like domains   24
low molecular mass organogelators (LMOGs)   136
lower critical solution temperature (LCST)   115
lysophospholipids   50

m

membranes   7, 8, 30, 31
- lipid monolayers
-- phospholipid monolayers, air/water interface   8–10
-- phospholipid monolayers, interfacial behavior   10–14
-- phospholipid monolayers, oil/water interface   10
-- phospholipid/protein composite layers, interfacial behavior   17–20
-- protein layers, oil/water interface   15–17
- modeling hydrolysis *in vitro*   20, 21
-- PLA_2 phospholipase   21–23
-- PLC phospholipase   23, 24
-- PLD phospholipase   24, 25
- polyelectrolyte-supported lipid bilayers   25–27
-- multilayers on curved surfaces   28–30
-- multilayers on planar surfaces   27, 28
mesenchymal stem cells (MSCs)   166, 167
microcapsules   41, 42
microcapsules, biomimetic   58
- applications   55
-- antibody adsorption   57
-- targeting   55–57

– ATP biosynthesis
– – proton gradient generation   76–78
– – proton gradients from GOD
     capsules   80–82
– – proton gradients from oxidative
     hydrolysis of glucoses   78–80
– biointerfacing   47, 48
– – asymmetric lipid bilayers on LbL-
     assembled capsules   51–53
– – asymmetric lipid bilayers on LbL-
     assembled protein capsules   53–55
– – lipid bilayer-modified   48–51
– layer-by-layer (LbL) assembly   42–45
– – permeation and mechanical
     properties   45–47
microtubules   92, 93
molecular biomimetics   41, 42
molecular motors   63, 64
Monte Carlo simulation   133
multivalent cationic lipopeptide
     (MCL)   142–144

n
nanofabrication   103
– heirarchy of dimensions   104
nanofibers   155
nanofibrils   136–139
nanoparticles   156–160, 161
nanotube arrays   131–136
nanotubes   131–136
nanowires   139, 140
Ni^{2+}-nitrilotriacetic acid (Ni-NTA)   67
4′-nitro-1,1′-biphenyl-4-thiol (NBT)   110
– SAMs   111
(N-(7-nitro-2,1,3-benzoxadiazol-4-yl)-
     phosphocholine)   49

p
Parkinson's disease   136
peptide encapsulated clusters (PECs)   164,
     165
peptide-based biomimetic materials   129,
     130, 171, 172
– applications   165
– – bioimaging   169, 170
– – biosensors   170
– – delivery of drugs or genes   167, 168
– – nonbiological applications   170, 171
– – three-dimensional cell culture
     scaffolds   165–167
– bottom-up nanostructure fabrication
– – amphiphilic peptides   150–156
– – aromatic dipeptides   130–140
– – lipopeptides   140–145
– – polypeptides   146–150
– peptide–inorganic hybrids   156
– – adaptive hybrid supramolecular
     networks   164, 165
– – inorganic nanoparticles on peptide-based
     scaffolds   156–160
– – peptide-based
     biomineralization   160–164
phosphatidylcholine lipids   55
phospholipid monolayers
– air/water interface   8–10
– interfacial behavior   10–14
– modeling hydrolysis *in vitro*   20–21
– – PLA_2 phospholipase   21–23
– – PLC phospholipase   23, 24
– – PLD phospholipase   24, 25
– oil/water interface   10
– phospholipid/protein composite
     layers   17, 18
– – dynamic adsorption mechanism   18,
     19
– – skin-like films on curved surfaces   19,
     20
– polyelectrolyte-supported lipid
     bilayers   25–27
– – multilayers on curved surfaces   28–30
– – multilayers on planar surfaces   27, 28
phosphotungstic acid (PTA)   164, 165
photosynthesis, artificial   74–76
PLA_1 phospholipase   20
PLA_2 phospholipase   20, 21–23, 50
PLC phospholipase   20, 23, 24
PLD phospholipase   20, 24, 25
polarization-modulated external IR reflection
     absorption spectroscopy (PM-IRRAS)   22,
     24
poly(acrylic acid) (PAA)
– PAH multilayers   44, 45
poly(allylamine hydrochloride) (PAH)
– PAA multilayers   44, 45
– PSS multilayers   44
poly(diallyldimethylammonium chloride)
     (PDDA)   51
poly(N-isopropylacrylamide)
     (PNIPAM)   115–118
– fabrication of complex brush
     gradients   119–123
poly(sodium   4-styrenesulfonate) (PSS)
– PAH multilayers   44
polyaniline (PANI) nanotubes   170, 171
polymer brush patterns for biomedical
     applications   114

– fabrication of complex brush gradients 118–123
– thermosensitve patterns for cell adhesion 114–118
polymerosomes 82–85
polyoxometalates (POMs) 164, 165
polypeptides 146–150
prion disorders 136
profile analysis tensiometry (PAT) technique
– experimental setup 11
– phospholipid layers
–– interfacial behavior 10, 11
–– oil/water interface 10
protein layers
– oil/water interface 15
–– kinetcis of protein adsorption 15, 16
–– skin-like films on curved surfaces 16, 17
– phospholipid/protein composite layers 17, 18
–– dynamic adsorption mechanism 18, 19
–– skin-like films on curved surfaces 19, 20
protein machines 91
proteins
– covalent immobilization 108–110
– electrostatic immobilization 105
–– lipid-modified HAS for *E. coli* recognition 107
–– lipid-modified HAS for targeted recognition 105–107
– LbL-assembled capsules 53–55
protein-transduction domains (PTDs) 168
proton pumps 71–74
proton-driven molecular motors 65, 66
pyranine 76, 77

q
quantum dots (QDs) 158–160
quartz crystal microbalance (QCM) 51, 149

r
ribbons 136–139
rubella-like particles (RLPs) 55, 56

s
self-assembled monolayers (SAMs) 110–112
– fabrication of complex brush gradients 119–123
silver nanocrystals 161
SNARE proteins 144, 145
storage modulus 136
supramolecular networks 164, 165
surface initiated atom transfer radical polymerization (SI-ATRP) 115
surface initiated polymerization (SIP) 114
surface pressure/area diagram of lipid monolayer 9, 12

t
thermogravimetric analysis (TGA) 138
three-dimensional cell culture scaffolds 165–167
transfer of nanocrystal organogel into water (TNOW) method 158, 159
tubulin 93

v
vesicles 131–136

x
X-ray photoelectron spectroscopy (XPS) 112

y
Young modulus 123, 131

z
zeta potential 43, 51